REIMAGINING THE
MORE-THAN-HUMAN CITY

Urban and Industrial Environments Series

Series editors: Robert Gottlieb, Henry R. Luce Professor of Urban and Environmental Policy, Occidental College

Bhavna Shamasunder, Associate Professor of Urban and Environmental Policy, Occidental College

REIMAGINING THE MORE-THAN-HUMAN CITY

STORIES FROM SINGAPORE

JAMIE WANG

THE MIT PRESS   CAMBRIDGE, MASSACHUSETTS   LONDON, ENGLAND

The MIT Press would like to thank the anonymous peer reviewers who provided comments on drafts of this book. The generous work of academic experts is essential for establishing the authority and quality of our publications. We acknowledge with gratitude the contributions of these otherwise uncredited readers.

This book was set in Stone Serif and Stone Sans by Westchester Publishing Services. Printed and bound in the United States of America.

Library of Congress Cataloging-in-Publication Data

Names: Wang, Jamie, author.
Title: Reimagining the more-than-human city : stories from Singapore /
    Jamie Wang.
Description: Cambridge, Massachusetts : The MIT Press, [2024] | Series: Urban and
    industrial environments series | Includes bibliographical references and index.
Identifiers: LCCN 2023058828 (print) | LCCN 2023058829 (ebook) |
    ISBN 9780262550932 (paperback) | ISBN 9780262381413 (epub) |
    ISBN 9780262381406 (pdf)
Subjects: LCSH: Urban ecology (Sociology)—Singapore. | Urbanization—
    Environmental aspects—Singapore. | Sustainable urban development—
    Singapore. | City planning—Environmental aspects—Singapore. | Climate change
    mitigation—Technological innovations—Singapore. | Climatic changes—
    Government policy—Singapore. | Environmental policy—Singapore. |
    Nature—Effect of human beings on—Singapore. | Singapore—Environmental
    conditions—21st century.
Classification: LCC HT243.S55 W46 2024 (print) | LCC HT243.S55 (ebook) |
    DDC 307.1/416095957—dc23/eng/20240221
LC record available at https://lccn.loc.gov/2023058828
LC ebook record available at https://lccn.loc.gov/2023058829

10 9 8 7 6 5 4 3 2

For my parents
who taught me how to imagine.

CONTENTS

ACKNOWLEDGMENTS

This book, and my interest in urban sustainability and technological sustainable solutions have grown from questions around justice, my ambivalence about the fascinating and troubled urban space, my anxiety concerning the inequalities magnified in unevenly developed and dense environments, and a sense of hope for other possible ways of inhabiting the world. I am truly grateful to have had the opportunity of thinking, researching, and reimagining a more-than-human world over the past years.

More importantly, this book would not have been written without the many who have helped and held me in myriad ways. Above all, my whole-hearted thanks to Thom van Dooren, my doctoral supervisor and cherished mentor, for his unfailing support, patience, guidance, and trust in this project since its inception, and for teaching me that storying and telling stories well is a deeply ethical work. I give my deepest thanks to Stephen Muecke, Astrida Neimanis, and Fiona Allon for their intellectual generosity and warm mentorship, and for showing me how to write and live with accountability and care.

I was enriched and benefited greatly from critical reflections and thinking together with the fellow researchers at UNSW and the University of Sydney as the book was taking its shape. My gratitude to everyone in the Environmental Humanities department at UNSW and the Department of

Gender and Cultural Studies at the University of Sydney. At the Sydney
Environment Institute and Sydney Southeast Asia Centre, I thank Danielle
Celermajer and Michele Ford for their inspiring advice. I am grateful to
Christian Winter, David Schlosberg, Natali Pearson, Genevieve Wright,
Ariane Defreine, and Suhasini Gunatillaka for having me at a range of
events the centers organized and their support.

Much of the revision of the book was also completed at the Education
University of Hong Kong. I am thankful to my wonderful colleagues and
to have this vibrant and rich intellectual home. Special thanks to John
Erni, Bidisha Banerjee, Jeffrey Clapp, Eric Yu, Kelly Tse, Hawk Chang, Zimu
Zhang, Man Lut Chau, and Karmen Zheng for their mentorship, insights,
support, and good laughs. I also extend my thanks to Mark Thompson and
Diego Fossati, and the Southeast Asia Research Centre at the City Univer-
sity of Hong Kong for first welcoming me to the research world in Hong
Kong.

As the book worked toward its final form, I was fortunate to be a visiting fellow and to present draft chapters on a range of occasions. I am
thankful for all the thoughtful and rich comments and suggestions from
the audience. My warm thanks to the Smart Sustainable Cities Research
Network and the Greenhouse Centre of Environmental Humanities at
the University of Stavanger, Norway, to Anders Riel Müller, Finn Arne Jørgensen and Dolly Jørgensen for their hospitality and the wonderful greenhouse members. I thank the Joint Centre for Advanced Studies: University
of Heidelberg and Rachel Carson Centre for Environment and Society for
the opportunity to think with inspiring groups of researchers. My sincere
thanks to Barbara Mittler, Christof Mauch for hosting me, and to Matthias
Schumann, Lena Engel, Katie Ritson, and Xiaojie Chang for your care and
joyful conversations.

I give my deep thanks to the fellow researchers for their support, feedback, and friendship along the way, and whose work keeps motivating me
including Hélène A. Le Deunff, Kate Judith, Sophie Chao, Blanche Verlie,
Sue Reid, Laura McLauchlan, Myles Oakey, Sam Widin, Stelle Wang, Emily
Zong, Alex Gearin, Eric Feng, Daren Leung, and Benjamin Iaquinto. I am
indebted to everyone who has discussed the key ideas and taken the time
to read the various iterations of the chapters. Particular thanks to Donna

Houston, Aidan Davison, May Ee Wong, Creighton Connolly, Lorraine Shannon, Chua Beng Huat, and Cheng Nien Yuan for their invaluable comments, and to Craig Santos Perez, Jennifer Wong, Mark Tredinnick, and Paola Caronni for their suggestions to the poems.

Various projects, readings groups, and collectives have nourished my thinking and the book over recent years as the world goes through overlapping crises, in particular the Learning Endings project with Astrida Neimanis, Patty Chang, and Aleksija Neimanis; At a loss for words of loss project with James Dunk, Freya MacDonald, Christine McFetridge, Cameron Muir, Anastasia Murney, Lynda Ng, and Kate Stevens, and the Multispecies Justice Collective. My editorial colleagues at the journal *Feminist Review* have become a source of strength where I draw courage and continue to learn to practice generous, attentive reading and thinking.

I must extend my warmest thanks to Emily Potter, who was instrumental in introducing me to the ethos of being a researcher. I also cherish the teaching from Marion Campbell, Ann McCulloch, and Josephine Scicluna at Deakin University, and Stephanie Han to relearn the art of writing. And special thanks to Gregory Morwood and Prudence Black for their kindness.

It is impossible to tell a meaningful story of urban development or to examine environmental issues without placing the more-than-human relations at the center. I learned this from literature but moreover from my field experience in Singapore. My profound and humble thanks to the many humans and other-than-humans who have guided me in diverse ways. My immense gratitude to the many people in Singapore for their time and generosity, to talk to me and show me how to observe, re-collect, and reimagine a capacious and generous world for the many beings in an increasingly dense, developed space and troubled times. I am inspired and will continue to be by their devotion, persistence, and creativity toward the important work they carry on.

The research, writing, and thinking were also made possible by a number of generous scholarships and grants from the Australian Postgraduate Awards, Research Training Program, UNSW top-up research scholarship, UNSW conference funding, University of Sydney Faculty of Arts and Social Sciences research and top-up scholarship, UNSW and University of Sydney Fieldwork Research Grant. I am thankful for Sydney Southeast Asia Centre awarding me its inaugural Writing Fellowship that allowed me further thinking space.

I thank the MIT Press for their trust in this project and for bringing this book to a wider world. My heartfelt thanks to my editors Beth Clevenger, Robert Gottlieb, and Anthony Zannino for their guidance and insights. I am grateful for Roger Wood of the MIT Press and Rashmi Malhotra of Westchester Publishing Services for guiding me through the production process. I extend my sincerest thanks to the anonymous reviewers for their generous and constructive feedback.

An earlier version of chapter 3 was published in *Cultural Studies Review* 25 (2), and an earlier version of chapter 5 was published at *Humanities* 10 (1), as part of Special Issue: Food Cultures & Critical Sustainability. Early versions of the poems *Who holds your Name* and the closing poem appeared in *Voice and Verse Poetry Magazine* and in "Urban": *A-Z of shadow places concepts* respectively. *The genealogy of tap water* was first published in *Otherwise Magazine*. I thank these journals and the editors for helping me to continue to shape the book.

Thanks to all my dearest friends and family members for their ongoing encouragement and understanding throughout the process. In particular thanks to Hui Yee Khaw for your unparalleled friendship and unwavering belief in this project and me. Your care for me over the decades knows no bounds. To my parents, I am forever indebted to your unconditional love and for holding me in all ways. This book is dedicated to you.

Finally, thanks to my partner Justin Wong. To list what you have done for me and the length you are prepared to go to support me would be an impossible task. May we continue to reimagine and discover wonders in the many possible futures.

INTRODUCTION: FABRICATING THE FUTURE

MAKING THE URBAN ENVIRONMENT IN CONTEMPORARY SINGAPORE

My eyes were certainly opened when I walked among hundreds of trade show booths at Singapore's 2017 Build Eco Xpo Asia and International Green Building Conference. People bustled in all directions throughout the multiple exhibition halls. Vendors, exhibitors, researchers, and representatives from various government agencies were all eager to showcase their new sustainable materials and software solutions for industrial, commercial, and domestic use. The sellers were animated as they demonstrated their cutting-edge technologies and sustainable products designed to make buildings greener and protect human well-being, and at the same time provide their users with a competitive edge. Visitors from diverse backgrounds looked equally enthusiastic. Here, sustainable and innovative environmental solutions to reduce urban heat island effects, improve building performance, and create energy-saving were inexorably aligned with the flow of capital while imbued with the sentiment of caring for the planet. Like many others, I was mesmerized by the pervasive atmosphere of hope. To construct a low-carbon future in tandem with economic prosperity was not only possible; it seemed imminent.

The annual International Green Building Conference is organized by the Building and Construction Authority of Singapore. It is one among an array of high-profile international events on sustainable and smart urban solutions that are now held in Singapore as part of the country's effort to position itself at the forefront of sustainable urban living. Marina Bay Sands, the venue for the convention, stands on some of the most expensive pieces of land in Singapore. It also houses the island's most iconic hotel and the city-state's only casino that, while controversial among its citizens, nevertheless attracts the wealth that the city so desperately needs to maintain its lifestyle (Ong 2016). Clearly, no expense or effort was spared on the convention, from the striking venue to an impressive lineup of speakers including policymakers, industrial leaders, and academics from around the world, all sharing their experiences of and research on environmental issues, the prospect of eco-cities, and the expanding market of sustainable urban solutions. Between the packed and tightly run program, business and government delegates enthusiastically networked and negotiated potential green deals.

Among all the green futures and products, I found myself particularly drawn to the various skyrise greenery solutions presented by commercial companies and government agencies such as the Housing and Development Board of Singapore. A wide range of modulated green walls or roofs planted with different types of vegetation were on display. In addition to offering a plethora of environmental benefits such as improving air quality and cooling the building, this vertical greenery highlighted attributes resembling types of do-it-yourself furniture, including easy installation, low maintenance, and the flexibility of mix and match. If skyrise greenery remains a stylistic novelty in many other parts of the world, in Singapore, it is a neatly assembled, packaged, and popularized marketable urban solution that supports greening while growing (see figure 0.1). It is also a central part of Singapore's vision of creating a vertical garden city.

Rated as one of the world's greenest and most sustainable cities, Singapore openly defies the stereotype of an urban concrete jungle.[1] In the midst of climate change and intense urbanization, cities are increasingly seen as "one of the most important solutions to these other grand challenges" (Christensen and Heise 2017, 436; see also Keil 2020). The development of sustainable and livable cities is gaining traction. In recent years,

0.1 A variety of skyrise greenery systems on display at HortPark, Singapore. *Source:* photo by author.

Singapore has been actively promoting itself as a sustainable urban model (Pow 2014; Hamnett and Yuen 2019). The *Sustainable Singapore Blueprint 2015* identified establishing the city-state as a global green hub to be a key component in the state's sustainable green agenda (Ministry of the Environment and Water Resources and Ministry of National Development 2014).[2] Not only are the buildings in the city-state replete with organic life but David Attenborough, the revered wildlife narrator, was also amazed by the abundant wildlife in Singapore. His charismatic voice in the documentary *Wild City*, directed by Claire Clements (2016), introduces families of otters, red junglefowls, and various kinds of birds in Singapore: their movements are carefully captured, dramatically juxtaposed with an intensely urbanized city and its stunning skyline. Sipping a Singapore sling at one of the many urban forest-themed rooftop restaurants, I recalled the convention hall at the Marina Bay Sands overflowing with people, capital, and promises of technological solutions for a better future. It became clear that the sustainable urban living that Singapore promotes is a particular kind of sustainability, paired with high-tech and persistent economic growth. As my time in Singapore continued, I discovered, in this seemingly harmonious techno-capitalist-green scene, that less visible but more complex stories also existed.

While I didn't have the pleasure of spotting playful otters during my time in Singapore, I came to learn about other types of wildlife including macaques, snakes, crows, and stray cats, those that, unlike the charismatic otters, are often unwelcome in the city.[3] For example, although a condominium sitting right at the edge of the forest is highly desirable, the macaques that share the same living space often are not.[4] Likewise, there is concern at the rising media coverage over resentment and anxiety toward the animals that are said to intrude on humans' space.[5] Culling is a likely outcome if residents file appropriate reports (Yeo and Neo 2010; Siau 2017).[6] The Agri-Food and Veterinary Authority of Singapore is reluctant to discuss such subjects or actions and often does not release the actual figures of culling.[7]

At the same time, the amount of roadkill has steadily risen over recent years as a result of further encroachment into the nature reserves.[8] In 2019, the *Straits Times* reported that the land in Singapore "changed from being a net absorber of carbon in 2012 to a net emitter in 2014, due largely to land conversion from forests and other vegetated areas to settlements" (Tan 2019).[9] Although Singapore has consistently positioned itself as a low-carbon-emission country, contributing only 0.11 percent of global emissions, as of 2017, this figure does not include the oil refineries operating on human-made Jurong Island as their products are moved and used elsewhere (Newman 2019). Similarly, marine bunker fuel was not included in the evaluation of per capita emissions because these emissions do not count toward Singapore's domestic figures (Schneider-Mayerson 2017).[10] As one of the world's largest ports in terms of cargo containers, if these emissions were included, Singapore's per capita emission level would be approximately five times higher than that of the United States (Ng 2012, 195). Seen in this light, envisioning a sustainable and livable city seems to be carefully crafted to exclude obvious associations with any ecological impact resulting from the country's economic engines.

In 2019, Singapore had one of the worst air quality days in years due to haze. The pollution level was so severe that the annual Formula 1 Grand

Prix, which draws large crowds of tourists and involves significant amounts of revenue, was threatened with cancellation (Sim 2019).[11] This seasonal haze has been a persistent hazard in an otherwise green haven.[12] The toxic fumes, which travel from neighboring Indonesia, are caused by the practice of forest clearance known as slash-and-burn where the land is set on fire to prepare for plantations, primarily oil palm.[13] While Singapore has outsourced the majority of its manufacturing arms, the domestic pollution level is constantly monitored and reported in the news to ensure the safety of its citizens. Yet the toxic haze—the product of both biomass burning and the greed of global capitalism's monocropping—is a reminder that the atmosphere does not understand boundaries between nations, domestic and public space, and rich and poor, nor does it abide by calculated figures of domestic emissions.

From the international conferences that promote green buildings, research, and trading of sustainable solutions to the city-state's carefully considered carbon emissions, from excluded urban wilds to the burning forests in neighboring countries, there are alarming signs of the unevenness of Singapore's progress, suggesting much more diverse, paradoxical storylines in the city-state's pursuit of urban sustainability and livability.

Indeed, urban environments offer a set of rich, complex, yet often compromised storied-places that are burdened with anthropogenic environmental damage and are themselves catalysts for many emerging environmental problems. This book explores how an alluring eco-futuristic urban imaginary is enacted and often driven by a specific, narrow version of sustainability in one of the most urbanized and modernized cities. How are diverse biotic and abiotic elements such as urban greenery and infrastructures shaped by, and in turn themselves shaping of, the nation and its economy and politics? What are the consequences, dangers, and ethical implications of enacting a particular urban imaginary that is beneficial only to select humans and nonhumans? As this book elucidates the flawed logic of some taken-for-granted urban sustainability practices or solutions in urban development, it asks what it might mean to reconfigure contemporary practices and ethics toward more multiplicitous ways of living in an increasingly urbanized environment. How might we think of urban relations differently if we take a broader more-than-human world seriously?

THE ISLAND AND URBAN IMAGINARIES

Singapore, an island city-polis in Southeast Asia, is a global financial center with a tropical climate and a multicultural population. With about 5.7 million inhabitants, as of 2019, living across approximately 720 square kilometers, the city-state is one of the most densely populated countries in the world (Singapore Department of Statistics 2019; The World Bank 2021). Known for its authoritarian one-party government, Singapore has been governed by the People's Action Party (PAP) since independence in the 1960s. Although much has been written about the country, the focus is often on its miraculous economic development from third world to first, the meticulous and paternalistic approach to governing and planning, the clean and green environment, and its ability to transform its vulnerability around such issues as water shortages into strengths. Singaporean scholar Harvey Neo describes the city-state as a developmental state, meaning that it "draws its legitimacy from the ability to continually develop and as long as the economy continues to thrive, the strength of this legitimacy is secure" (2007, 197; see Pow 2009). In recent years, Singapore's rising profile as a sustainable city able to maintain its economic growth has given rise to a proliferation of scholarly work and media reports on Singapore as an experimental city, a sustainable city, and a smart city.[14]

While city managers and national governments in Asia and beyond express substantial interest in Singaporean planning practices, Singapore is actively commodifying its urban solutions: from exporting the Singapore standard for green buildings to its infrastructure and approaches to water and urban mobility (Dale 2008). Furthermore, it has packaged its technological urban solutions, sustainable narrative, and urban planning policies through government consultancies to build eco-cities in other countries (Pow and Neo 2015: Caprotti, Springer, and Harmer 2015). For example, Sino-Singapore Tianjin Eco-City (SSTEC) is a prominent state collaborative project in China's recent environmental turn. The project showcases the city-state's strengths in housing, environmental services, and water technology. The agenda animating this project is that SSTEC might be a prototype city built on Singapore's "distilled" knowledge of urban solutions that can be replicated in other parts of China (Ministry of National Development 2018).

In the era of the Anthropocene, the proposed new geological epoch in which humans are seen as a significant force shaping the planet, many, in particular the socially, economically, and geographically disadvantaged, are losing control of everyday life at an alarming speed as they face unprecedented extreme weather events and environmental issues.[15] As Rob Nixon (2019) says, "We may all be in the Anthropocene but we're not all in it in the same way." Environmental humanities scholar Matthew Schneider-Mayerson (2017) explores precisely this inequity in his work on Singapore. As he hauntingly points out, although many cultural imaginings of climate change point toward a dystopian future of sea level rise, and although low-lying islands are particularly vulnerable, Singapore, one of the world's richest countries, has the wealth to engineer a way out. While others sink, Singapore will rise. At least that is the plan.

In exploring these contemporary and complex challenges, this book tells the story of pressing, distinct, and intra-related environmental issues in Singapore and the resultant multifaceted ecological challenges, from urban greenery to housing development projects, transportation, water infrastructure, and urban farming. Some topics explored in the book, such as Singapore as a "garden city" and its water technology, have been widely documented yet often approached from a human-centric governance and policy level. As opposed to many scholarly works on Singapore, this book is not a textual analysis of Singapore's urban policies, nor does it offer a comprehensive overview of urban environmental issues.[16] Instead, drawing on my fieldwork in Singapore, including interviews, observations, participation, and personal experiences, this book offers a detailed, contextualized way of thinking about the making of the urban environment, guided by the storylines of humans and nonhumans that are woven together in urban life.[17] With Singapore's positioning as a model of sustainable urban living, and given the overwhelming dominance of the official narrative, it is vital to foreground stories that attend to differences and multiplicitous modes of life in order to make visible the obscured "we" in the Anthropocene. At the same time, the book highlights emerging responses to changing social, economic, and environmental dynamics such as the rise of urban farming. Importantly, in line with a significant body of scholarship, this book refuses the notion of a "nature" that is distinct from human life and instead explores the multiple ways in which natures are imagined and co-constructed by humans and other-than-humans.[18]

In environmental humanities scholarship, *storying* is understood as a rigorous method, a dynamic and "slow, careful, work" of attending to the world (van Dooren 2020a, 3; see also van Dooren and Rose 2016). As anthropologist Deborah Bird Rose points out, "Stories themselves have the potential to promote understandings of embodied, relational, contingent ethics" (2013, 9). In this book, storying is also a mode of inquiry into imagining and reimagining, drawing us into ways of creating possible futures and relations of humans and nonhumans.

In recent years, there has been a small but growing body of work on urban imaginaries. For example, Christoph Lindner and Miriam Meissner highlight the critical role of urban imaginaries "in shaping the future of urban societies, communities, built environments, and ecologies" and the need to flesh out the *"politics of urban imagination"* (2019, 2 and 12, italics in original; see also Çınar and Bender 2007). Imagining is also an effort of world-making. As Emily O'Gorman explains, *imaginings* do not exist independently of materiality, rather they "are always situated and partial" and "guide actions and in so doing help to shape particular kinds of worlds" (2017, 3).[19]

In this book, I focus on the various ways in which Singapore has been imagined since its independence in the 1960s and explore how these imaginaries have been enacted and incorporated into urban narratives, and their social, cultural, and political implications. For example, this book examines the process through which Singapore's visioning of a clean, green "garden city" has evolved over the years that has shaped, advanced, and sustained a variety of economic and development-focused agendas. Most recently, the city-state's march toward progress coupled with the advancement of technology has given rise to a new kind of imagining that goes by the name of ecological modernization. The central tenet and promise of this theory is that of "overcoming the environmental crisis without leaving the path of modernization" (Spaargaren and Mol 1992, 334). This is one of the key theories that I discuss further in the following sections.

Thinking with and furthering these discussions on urban imaginaries, this book explores how we might develop ways of re-thinking, re-seeing, and re-storying cities. By understanding the work of imagining as a highly consequential material and semiotic practice, it further draws out the disavowal, separation, and yearnings that emerge in the multifaceted and contested narratives of urban development. Reimagining is also creative, gesturing to alternative ways to design or inhabit environments, to reconnect, repair, and craft other possibilities of world-making. In addition to demonstrating that the urban is an important site for addressing pressing environmental challenges, this book demonstrates how (re)imagining the urban helps to develop other framings, approaches, and responses to some of the many key conceptual issues that concern accelerating environmental issues such as questions of coexistence, mastery, displacement, and justice.

In addressing these challenging and interrelated questions, this book brings three main strands of critical thinking within the environmental humanities and social sciences into conversation around urban life. These are theories of ecological modernization, of more-than-human communities, and of the notion of future and futuring. In contributing to the debates and development in these three thematic areas, this book does not limit itself to a single discipline such as urban planning, cultural geography, environmental philosophy, or political ecology. Rather it places the exceedingly influential narrative of modernization in the broader interdisciplinary field of the environmental humanities, to open it up to further questions and alternative solutions.

As Singapore has become an emblematic New World city, its mode of urban sustainability has both broad and deep implications for many cities around the world that are eager to become greener and more sustainable. This also includes the emerging and rapidly growing urban areas and the dramatic rural-to-urban transitions that are underway in many parts of the developing world (Tan and Abdul Hamid 2014). Ultimately, by attending to humans and other-than-humans that co-shape the city-state, this book makes a critical intervention into human-centered, technocratic, capitalist modes of urban development, and advocates a version of relational and careful imagining to open up a space for more diverse and inclusive futures.

ECO-MODERNIZATION

Environmental concerns are attracting increasing attention globally, and, as a consequence, modernization's co-opting of the desire for an eco-future is morphing into eco-modernism. This transition is exemplified by Singapore's more recent pursuit of prestige as a leading livable and sustainable city of the future. Unlike some of the many pressing images that capture urban environmental degradation, be it the heavy smog in Delhi, severe drought in many parts of Australia, or sinking islands in the Pacific due to sea-level rise, Singapore offers a glowing picture of an inspirational green futuristic city where the impacts of climate change appear to be actively considered and incorporated into policy and practice.

The rise of ecological modernization theory in the 1980s started as a response and challenge to environmental movements and the idea that sustainability requires deindustrialization, reduced economic growth and consumption, and simpler lifestyles (Mol 1995, 2002; Hajer 1997).[20] In the eco-modernist account, "further industrialization and technological innovation," including clean technologies and "new processes to monitor the environmental impact of production and consumption," provide a way out of ecological crisis (Wong 2012, 99).[21]

The premise of coupling ecology and modernization gels with the modern pairing of sustainability and continuous development. First proposed in *Our Common Future* (WCED 1987), sustainable development is defined as "development that meets the needs of the present without compromising the ability of future generations to meet their own needs." If sustainable development aims to maintain the status quo, eco-modernization seems to offer the means to take it further. Not only does it encourage business as usual but also the discourse "is precisely about trying to represent environmental issues as profitable enterprise" (Harvey 1993, 4–5).[22] In Singapore, the direction of the environmental governance policy championing green growth through technology and state planning is aligned tightly with ecological modernization (Wong 2012). Here, the aim is to turn vulnerability into opportunity; technological development, economy, control and environmental issues complement and reinforce each other (MEWR and MND 2014). Being green, if planned and executed properly—as seen in the bustling trade exhibition at the conference—enhances the economic future.

An unwillingness to slow down the modernization process and hence the inability to imagine a significantly different path to the future has led to some "creative" proposals. In 2015, "An Ecomodernist Manifesto" was released by a collective of international, mostly US-based, environmental thinkers (Asafu-Adjaye et al.). Using intentionally provocative language, the manifesto celebrates the Anthropocene as the era in which humanity is taking full control. According to these eco-modernists, accelerated technological progress can allow humanity to mitigate climate change, spare nature, and relieve poverty (Asafu-Adjaye et al.). Hence, many human activities from desalination to nuclear power and industrial farming should become more intensified (Asafu-Adjaye et al.). These types of technological intensifications that are rooted in a disregard for ecological limits—at least in the sense that they might readily be overcome—is at the center of some of the discussions in this book.

Furthermore, this book demonstrates the limitations and implications of eco-modernization, in particular the wishful thinking of being able to alleviate humans from the mess we have created without radical changes in other social, economic, and political spheres. It contributes to critical discussions on eco-modernization by proposing two key points. First, I examine how the intensification underpinned by eco-modernization exacerbates many environmental issues. In the midst of unprecedented environmental challenges and intense urbanization, it is imperative to examine the continuing reliance on technology to curb environmental damage while seeking to overcome ecological limits. In the case of Singapore, a prominent example is the city's growing dependence on technological water production including seawater desalination and purified recycled water (chapter 4). This water technology has been commended as a significant achievement that not only solves Singapore's lack of water resources but also serves as an example of improving economic growth, job opportunities, and knowledge exportation. But there is little sign of abating water consumption, particularly its heavy industrial use. At the same time, the production of technological water is energy and capital intensive and produces significant CO_2 emissions. In this light, an eco-modernist optimism of water provision through techno-capital-centric solutions becomes a cruel attachment that ultimately obstructs one's own flourishing (Berlant 2011).[23]

Second, the politics of an alluring eco-modernist narrative that fore-sees no need for structural change displaces or distracts from many other important social and environmental impacts. Indeed, environmental or sustainable issues have often been framed as carbon-emission issues that can be solved or reduced by technologies (see Hultman 2013, 87) or man-aged through tax policy.[24] The simplified logic of low energy emissions equating to sustainability ignores other tensions and environmental issues in the urban environment. For instance, when sustainable transportation is narrowly framed as a task of reducing carbon emissions, all the related environmental consequences of these purported sustainable movements are portrayed as either necessary costs or able to be mitigated through sci-ence or a kind of disavowal that places things in the shadows (chapter 3). During the National Day Rally 2019, the prime minister of Singapore, Lee Hsien Loong, highlighted climate change as the island's main threat that must be treated with utter seriousness. He referenced the introduction of a carbon tax as a key measure. At the same time, policy in Singapore primar-ily focuses on improving efficiency through the advancement of techno-logical solutions. Yet, as Singapore continues to welcome the expansion of oil refineries and other industrial-scale players, curbing emissions through a carbon tax or improving efficiency addresses neither the root cause that produces these emissions nor other consequences.

Importantly, eco-modernization has also been used to soothe anxiety over the climate crisis. As ecological challenges are seen to be ultimately technological challenges (Asafu-Adjaye et al. 2015), neither industry nor the populace are compelled to introduce structural change or adapt their lifestyles. This is despite the fact that the affluent city-island-state is known for its materialist consumer culture with a high consumption of water and energy.[25] Singapore is certainly not alone on this aspect. In a comparable vein, anthropologist Gökçe Günel points out that the focus on renewable energy and clean technology in the development of Masdar—a supposed zero-carbon and eco-futuristic city—is never to challenge the status quo, but rather the goal is always to preserve the present "during a time of ecological destruction" through "technological innovations" (2019, 13).

In Singapore and around the world, the possibility of building a sus-tainable and livable city using technological advances without sacrificing economic development has been visualized as the way to move forward.

Meanwhile, others have flagged issues that have arisen in countries that adopted an ecological modernization theory. Rolf Lidskog and Ingemar Elander point out that Sweden, an early adopter of ecological modernization and a long-term champion of its efforts in sustainability, has never developed an environmental policy that challenges the logic of growth (2012, 421). Rather the country has been transferring its ecological footprint beyond its borders in a range of ways. With the rise of eco-cities or zero-carbon cities, some researchers are questioning the social success of these urban environments and describing them "as islands of wealth in an ocean of poverty" or built in the middle of industrial zones, suggesting they themselves are already, or will become, sites for inequality (Sharifi 2016, 11; Caprotti 2014).[26]

As eco-feminist Greta Gaard (2001) long ago said, the problems of environmental destruction and social injustice are bound up with each other. This book is deeply interested in understanding how humans and technology might be part of possible solutions, but it is equally concerned with the dangers of the ways in which (some) humans exert themselves on others and a broader environment through technology. The insatiable drive for progress disguised within the narrative of eco-modernism is enabled by technological advancement, neo-liberalism, and growing uncertainties posed by climate change. In the shadow of eco-modernization, it is crucial to ask how capital has been accumulated to sponsor specific modes of living while ignoring environmental limits.[27] How might some techno-centric solutions themselves aggravate the environmental crisis?

MORE-THAN-HUMAN WORLD

Compared to many other parts of the world, incorporating greenery into the urban environment has been an essential part of life in Singapore. Lee Kuan Yew, the city-state's founding father and its first prime minister, early on sensed the importance of nature in nation building. In his words, "A blighted urban jungle of concrete destroys the human spirit" (1995, 1). Moreover, Lee was one of the first to envision the link between attracting foreign capital and a well-manicured clean, green environment.[28] In Lee's account, a garden also helps to raise the morale and discipline of the citizens (2000, 175). Although the city-state has been widely commended for its inclusion of nature in its dense city environment, an ethos of control or instrumentalization is central to its approach to both humans and nonhumans. This book is concerned with these paradoxical approaches of foregrounding and diminishing natures and examines the issues of an anthropocentric mode of engaging nonhumans to pursue urban sustainability.

Nevertheless, for much of human history, the urban has been seen as an exclusively human space that excluded nature (Wolch 2002; Gandy 2006). Yet, William Cronon (1995b), in his discussion of the indispensable relations between city, country, and nature, points out that the production of the city in fact results in the continuous production of new urban "natures," of new urban social and physical environmental conditions. Over the past decades, and in particular since the 1990s, scholars from a range of disciplines have interrogated a plethora of nature-culture, nature-society, human, and other-than-human binaries, seeking to re-learn the interrelations and intra-relations between humans, nature, and culture (Haraway 1985; Plumwood 1993; Castree 1995; Latour 2004; Braun 2005). Cities are increasingly emerging as key sites to explore these entangled and conflicting relations.

In a similar vein, urban political ecologists have productively highlighted how the political process has informed the environment and the ways in which the social mobilization of the transfer of energy and resources produces particular socio-environmental urban conditions (Keil 2003, 2005; Swyngedouw and Heynen 2003). How has nature become integral to various forms of capital accumulation (Gandy 2003)? Yet, while these discussions "are enormously productive intellectually," Jon Christensen and Ursula K. Heise argue that "there is very little 'ecology' in political ecology" (2017, 439).[29] In agreement with Christensen and Heise, I am drawn to urban geographer Jennifer Wolch's stance that the literature on cities largely focuses on "habits of capital, class struggles, racial formations, and gender divisions, regulated by urban growth machines and their shadow state apparatchiks" and needs to do more to foreground the nonhuman actors such as "urban rivers or oak trees or red-legged frogs" (2007, 373).

As a highly compressed and dynamic environment, the urban is, in many ways, a vital and hopeful site for serious discussions around engaging with, or fleshing out, tensions between humans and nonhumans.

Lesley Green, in following stories of the Cape Town baboon, concludes that "In urban conservation, there are no wild spaces that are free of humans" (2020, 170). In a similar vein, the drastically different treatment of the urban wild in Singapore—from culling the ones deemed unwanted to championing those seen as a success in the city's sustainable work—has made clear the politics of inclusion and exclusion at work and requires deeper reflection on the disputes over urban natures.[30] Or as Sarah Whatmore argues, it is to initiate *"more-than-human* modes of enquiry"* (2006, 604, italics in original).[31]

The term "more-than-human world" is widely considered to be introduced by David Abram (1997), who, in thinking with indigenous cosmologies, calls for humans' re-immersion in a more-than-human sensuous realm. In recent years, the concept of more-than-human has been productively engaged with in a range of contexts in rethinking the shifting relationships among and between humans and nonhumans (Whatmore 2006; Rose et al. 2012; Probyn 2014; Chao 2022). In this book, the more-than-human world encompasses diverse *relational* entities from urban greenery to technological water, rivers and dams, wildlife bridges, exhumed tombs, and more. My analysis of the urban through a more-than-human approach examines and foregrounds the aspects of co-constitution and multiplicity (O'Gorman and Gaynor 2020; van Dooren and Rose 2016; Franklin 2017). I am interested in the kinds of engagements with, or attempted separations from, the more-than-human world that constitute the dominant imaginary of the urban world. How do some bio-social-cultural relations and understandings of more-than-human worlds open up possibilities, and not others, and with what consequences for whom?

In each chapter, exploring distinct and intra-related urban environmental issues through a more-than-human lens allows me to examine intra-human, and human and other-than-human relations in the development of contemporary Singapore, and the co-constitution of particular eco-futurist imaginaries. Reimagining a more-than-human city advocates the need to subvert human exceptionalism and "urban exceptionalism" that "encourages modes of thought that regard cities as places that have somehow risen above the physical constraints of 'nature'" (Houston et al. 2018, 192). It is also, in ecofeminist philosopher Val Plumwood's words, to resituate "humans in ecological terms" and "non-humans in ethical" and cultural terms (2002, 8–9; see also Plumwood 2012). Crucially, more-than-human or multispecies studies do not flatten the complexities of intra-human relations. A significant portion of Plumwood's work explores and demonstrates how modernity has imagined and worked to create human mastery over nature. This book aims to tease out and thicken the understanding of how that mastery is tangled up with other forms of intra-human domination. For example, many aspects of Singapore, from infrastructure and transportation to diverse types of care work, have been outsourced to foreign workers, of which many are from its neighboring countries (Wee, Lam and Yeoh 2022; Barr 2019). This book explores these layered and entwined forms of mastery in the urban context (chapter 1).

As Singapore is a highly modernized and urbanized place in the world that nevertheless imagines itself as a leading sustainable and green city, this book shows how urban natures in the city-state are simultaneously foregrounded and disavowed. It argues that acknowledging the role and importance of nonhumans in the city-state is essential, as is appreciating cities as part of a wider environment. Yet, these approaches might not be sufficient to move away from the pull of an eco-modernist future that centers on an instrumentalizing, anthropocentric, and increasingly "creative" way of approaching natures. The recognition of our reliance and situatedness in nature does not necessarily evoke an ecological way of caring; it may, in fact, increase a sense of vulnerability and exacerbate the desire to develop solutions that appear to counteract the uncertainty of nature (chapter 4).

Some researchers regard Singapore's green image as having been fully commodified (Gulsrud and Ooi 2015). For example, a charismatic or techno-hybrid nature has been mobilized to produce an image of future living and draw large numbers of tourists (chapter 1). In response to these discussions, this book raises critical awareness of the particular urban natures imagined and produced in Singapore that are inseparable from the broader eco-modernizing agenda of the city. These are deployed to satisfy both the needs of development and a particular type of sustainability. Singapore's tech-supremacy also results in further control of more-than-human relations. This includes eco-modernist research to develop "climate-resistant" technological water or food (chapters 4 and 5). A more-than-human city may well include technologies, but in Singapore this is achieved in a specific way that is also inflected with its authoritarian style and a particular way of imagining an urban future.

At the same time, in reimagining a more relational city, throughout this book, I highlight the importance of attending to the uncontrollable aspects of more-than-human worlds as a way to subvert the dominant narrative and approach. For example, the severe erosion resulting from land reclamation has necessitated a reexamination of the crucial role of mangrove forests in ecosystems, from supporting biodiversity to safeguarding coastal areas. It also seeks to draw attention to the importance of respecting and acknowledging uncertainty and alternative modes of knowing in an urban environment (Whatmore 2006, see also Steele, Wiesel, and Maller 2019).

As David Attenborough (BBC 2016) takes the viewer around Singapore admiring the wildlife in the aforementioned documentary, he comments, "Create the space and the animals will come." In an effort to create space for natures, how, then, does the city along with its human residents prepare for coexistence? How might our way of interacting with nonhumans allow for more spontaneous growth? As natures are increasingly incorporated and integrated in the urban environment, there are many more complicated and emerging issues to be considered in urban planning than simply incorporating and sustaining various forms of "green." Here, I am referencing a growing body of literature on the need to cultivate livelier, affective, and convivial democratic relations in urban environments (Hinchliffe et al. 2005; Hinchliffe and Whatmore 2006; Houston et al. 2016). In exploring some of the stories that seek to create new possibilities of co-living in the urban environment, this book challenges the view of harmonious living underpinned by control. Instead, it argues for the need for more interconnected ways of coexisting that respect spatial and temporal connections. It is about how to live with others, even if sometimes uncomfortably, and how to understand their presence.[32] Multiplicity seen through a more-than-human lens, as we will see, is not imposed synchronization or forced harmony; rather it anticipates and prepares for conflicts and respects the prospect of living uncomfortably in an increasingly cohabited space.

FUTURE AND FUTURING

Singapore is widely considered to be a significant model for the future of urbanism, one in which the flourishing of the human is in harmony with the flourishing of nature alongside a vibrant economy. CNN's report series "Future Cities" focuses on the island-nation's exceptional forward-looking approach, from using underground space, to embracing skyrise greenery, to an integrated transport system, and more. Moreover, the series notes, "Singapore is now exporting its expertise in urban planning to other cities in Asia where rapid urbanization is taking place—including the Tianjin Eco city, China and the new capital city in Andhra Pradesh, India—and paving the way for cities worldwide to ensure they build sustainably and improve their liveability" (Senthilingam 2015). Sociologists Barbara Adam and Chris Groves suggest that as futures are emptied or decontextualized, they are easier to tame, transform, and traverse (2007, 2). In the green trade show held at the Marina Bay Sands, it was not only green technologies and products that were on sale but the promise of a more sustainable, comfortable, and profitable future.

For some, stories of the future start with Paul Klee's painting *Angelus Novus* (1920). Walter Benjamin offers his reflections on the angel:

This is how one pictures the angel of history. His face is turned toward the past. Where we perceive a chain of events, he sees one single catastrophe which keeps piling wreckage upon wreckage and hurls it in front of his feet. The angel would like to stay, awaken the dead, and make whole what has been smashed. But a storm is blowing from Paradise; it has got caught in his wings with such violence that the angel can no longer close them. This storm irresistibly propels him into the future, to which his back is turned, while the pile of debris before him grows skyward. This storm is what we call progress. (1968, 257–258)

Thinking with Benjamin, Rose (2006) describes the process of pressing forward to create a modernist and settler future as an ecological violence that ruptures the generational connectivity between life and death.[33] Consequently, this not only depletes the present but ultimately erases the past and future of beings, human and not. As (eco)modernism focuses on a linear green capitalist way of creating a sustainable and livable urban city, we must ask: Which future and past, and whose futures and pasts, have been foreclosed or disconnected?

Since its independence, Singapore and its leaders have looked toward the future. Cultural theorist Belinda Yuen encapsulates it succinctly, stating, "In its vision to become a world-class city, Singapore has turned urban development into a future-making project, a 'growth machine' for export, and portrayed itself as a promoter of culture, environment and quality of life for its residents" (2011, 155). In many government institutions, futures have already been meticulously planned, mapped out, and exhibited in the galleries of government agencies. Indeed, the planning system that maps out the island's future in great detail is seen as the cornerstone of the People's Action Party and is central to Singapore's successful development (Hamnett and Yuen 2019). But for many of the conservationists and other local people I spoke to in Singapore, the future seemed to be less certain.

Meanwhile, the impact of urban imaginaries is not only on the future but also on the present and the past. Among various planning schemes, a notable goal has been the active pursuit of population growth.[34] This is positioned as central to ensuring a sustainable economy. Consequently, it sparked another round of island-wide development works that has seen the expansion of public transport including excavating underneath a natural reserve, exhumation of tombs to make way for development, and the erection of sustainable eco-living housing estates that replace forests formerly inhabited by rich nonhuman lives. During this process, the pasts, the presents, and the futures of diverse human and other-than-human communities have been sacrificed for an imagined demand for infrastructure, housing, or parks among other projects, all in the name of building a particular sustainable future of continuous economic growth, based on technological advancement and a supervised population mix.[35] As land is appropriated for future use at the expense of the present lives of the many beings, the state colonizes not only space but also the temporal relations of diverse others. Here, a more-than-human inquiry into the future helps to further draw out the increasingly eroding relations between the living and the dead, and its continuous impacts on landscape and memoryscape (chapter 2).

Benjamin's allegory of the angel of history continues to haunt the eco-modernist vision of the future. As Bronislaw Szerszynski notes, the angel of history depicts the eco-modernizers' story underpinned by "technophilia and a profound belief in progress . . . accelerating through disaster and leaving it behind" (2015, 241). As I explore the urban farming that aims to be pest-free, soil-free, or dust-free, I am concerned by the common response to uncertainty about the future as a result of climate change; a response that seeks to cultivate a sense of invulnerability accompanied by a growing desire to hyper-control and develop ways of sheltering some humans and nonhumans. How do some of these projects, while attempting to provide for the future, simultaneously stymie or hinder the possibility of other kinds of futures, and with them the diversity that might ultimately be needed to adapt to different futures?

It is important to note that the various effects of environmental damage have their own temporalities. Some are imminent and forceful, while others, such as radioactive waste that takes ten thousand years for its radioactivity to fade to safe levels, can be slow or latent in their insidious effects. Bruce Braun, drawing on Bruno Latour, suggests that the pile of wreckage in the allegory of the angel of history is not the crumpled past that has been left behind, but rather it is the "looming catastrophe," the shape of things to come (2015, 239). In the context of Singapore and many other wealthy countries that may have exported their environmental impacts in different ways, it is particularly important to think with the temporal flow of this damage. For some of us who have contributed to the escalating effects of climate change in the process of anticipating a desired future at the expense of the present lives of others (Adam and Groves 2007), the rise of the island (Schneider-Mayerson 2017), perhaps for a short while, is ultimately an illusion. In this book, this type of temporal and spatial entanglement of environmental damage is examined in the discussion of some technological solutions and contrasted with more relational ways of imagining and worlding. For instance, is it possible for some forms of urban farming to cultivate futures that engender life and care?

In its focus on a highly modernized, futurist city, this book explores different modes of futuring, examining, on the one hand, "who is empowered and authorized to imagine and define socioecological futures" (Braun 2015, 240). And on the other hand, it invites a much more fundamental engagement with a multiplicity of ways of living in order to destabilize a singular notion of progress and security and open up space for possible more-than-human futures. In Arjun Appadurai's discussion on futurity, he advocates for a mode of futuring that counters an "ethics of probability" inscribed by "technocratic enlightenment" or nature's oscillations, one that takes seriously the future being a cultural fact (2013, 299). In this book, I push this line of inquiry further, attending to some more contingent yet hopeful future-making alternatives that seek to reconfigure and design contemporary approaches centering on more-than-human relations and "naturalcultural" scape (Haraway 2008; Fuentes 2010). These include building ecological corridors, re-naturalizing canals, and growing edible gardens in the cracks of the city.

In recent literature on the future, scholars are increasingly advocating for futures that give and support "diverse and autonomous forms of life and ways of living together" (Collard, Dempsey, and Sundberg 2015a, 323). Céline Granjou and Juan Francisco Salazar emphasize "the future-making potentialities of nonhumans where multispecies entanglements have critical consequences for acting and living in Anthropocene times" (2016, 242). Indeed, reimagining a more-than-human world in the urban environment requires rethinking how we engage with the future. There is a need to understand, for instance, that infrastructures are more than a technology of spatial connectivity for humans. Rather they have the ability to either offer or close off temporal possibilities that sustain relations for the human and other-than-human community (chapter 3).

In short, ecologically oriented urban stories must pay attention to messier, more complicated, and uncomfortable relations to be bound by care and cultivate a strong sense of commitment to sustaining not one kind of future, for one kind of being, but multiple and diverse futures (Goodyear-Ka'ōpua 2017). In thinking of which stories to tell, in whose language, and what connections they may enable or disable, we are ultimately questioning what type of future we are imagining, and with whom we are cohabiting. Coexisting well with others in the present is not only a question of acknowledging a shared rooted past but also one of imagining and working toward shared futures.

WEAVING THE STORIES

As the stories of this book move between the past, the present, and the imagined futures of Singapore, the three themes of eco-modernization, the more-than-human world, and futuring reverberate, refract, and interact, revealing the complexities, challenges, and potentialities of the vision of building a sustainable and livable city.

In addition to drawing on diverse scholarly work on ecology, conservation, urban planning, cultural studies, environmental humanities, and policy documents, newspapers, and other media sources, the reimagining work in this book and my analysis are grounded in the findings from multiple field trips to Singapore between 2017 and 2018. Along the way, I spoke with conservationists, policymakers, farmers, commercial managers, and a range of researchers in relevant fields including biology, environmental studies, social science, and urban solutions as well as a variety of local residents. I sought to collect insights, ideas, and impressions about different facets of the development of Singapore, urban natures, and their collaborative and/or contested relationship with environmental issues. What are the stories, relations, and interactions with specific landscapes? What are the possibilities that might have been opened up and closed down through some eco-modernist initiatives?

Alongside interviews, there are important inputs from site explorations and observations. During my time in Singapore, I visited a variety of urban eco-projects, nature reserves and parks, government galleries, a water production center and dams, a range of markets and urban farms, and many more. These allowed me some immersive and situated experiences of living in "a city in a Garden" and enabled encounters with more-than-human communities. In many ways, I have approached the field "as a dynamic and shifting entanglement of relations, rather than a property of things" (Barad 2007, 224). I traced the lives and journeys of heritage trees and transplanted trees. I commuted between sites with different modes of transport, noting the experiential differences. I walked in the cemeteries where graves continue to be exhumed to make way for development, revealing the multifaceted implications of the displacement and relocation of humans and other-than-humans. I visited the Kampong (meaning village in Malay), where local families helped me to discover the unsayable sticky relations moored in the landscape and the difficulties and hopes associated with living in a highly urbanized state.

As well as drawing on an array of policies and planning documents on Singapore's approach to urban sustainable solutions, I learned the official narrative firsthand through several tours offered to visitors by government agencies. At the same time, despite the dominance of the state's imagining and storytelling work in the life of this city, I was encouraged by a number of extremely well-organized volunteer-run programs that seek to make the public aware of the tensions of urban development, heritage, and wildlife preservation, and offer an affective, empathic, and rich introduction to Singapore's naturalcultural scape.

As Tim Ingold and Jo Lee Vergunst point out, walking "is itself a way of thinking and of feeling" (2008, 2). In my project, this practice of knowing (Ingold 2010) also attends to the modes of movement of other-than-humans, and is itself a process of world-making (Barad 2007).[36] In one case, it involved my crawling in the bushes imagining being a civet looking for a path to cross a fragmented nature reserve. Such explorations form a part of the approach that orients the storylines of the book (particularly in chapters 3 and 4). At times during the research, new voices or unexpected encounters seemed to carve out their own paths, alerting me to an overlooked part of the story and guiding my thinking.

In this book, storying is not a work of stitching things together. Rather, I bring interviews, observations, personal experiences, and diverse literatures into conversation. I draw on my findings selectively to elucidate particular insights of informants and to invite the reader to think about how they matter differently. As I think with the more-than-human relations situated in various urban sites and with their complicated issues and consequences, storying is also a crucial method that allows me to layer in different understandings and values, and weave together diverse narratives while holding onto "multiplicity and complexity" (Griffiths 2007). In other words, rather than aiming to produce an objective account, this book tells my own stories that account for and work with others' thinking without attempting a singular resolution.

More-than-human storying is always a dynamic process of unknowing and opening up to new modes of knowing. At the same time, these kinds of descriptions and recounting draw their valence from specificity. For example, throughout the book, I seek to highlight and draw out the empirical nuances of some human differences, inequality, and hierarchy in the pursuit of particular versions of sustainable urban futures and in their shaping of urban environments while resisting abstractions. In this context, the discussion of race, ethnicity, and migration is informed by a large body of literature that centers on these issues in various aspects of Singapore.[37] More importantly, my thinking on these topics is guided by and situated in the environmental issues explored in the book, such as the interconnectedness of varied forms of human and nonhuman oppression.

In the face of growing enthusiasm for building sustainable cities, what does Singapore's state-sponsored model of eco-modernism offer and affirm to other cities around the world? As one of the most urbanized metropolises globally, Singapore offers a position from which to rethink the importance of situating humans and cities in an ecological system. This includes examining the challenges and compromises that may come with it, as well as the danger of continuing the progress-driven and human-centric mode of its urban sustainable narrative.

In exploring ecological connections, this book does not advocate a kind of romanticism, or return to more "pristine" relationships with an exteriorized nature or an imagined ideal past. Instead, my exploration is firmly grounded in an ambivalent hope about technologically mediated paths. These contextualized analyses do not denounce techno-mediated solutions. As Donna Haraway writes, "It is neither technophobic, nor technophilic, but about trying to inquire critically into the worldliness of technoscience" (2004, 326). It needs to be noted that the practices and approaches in relation to some of the alternative nature-solutions that I explore are not total solutions to all the problems. Yet, these practices, coming with their own compromises, do enable different possibilities. Through these empirically grounded analyses, my aim is to bring halting voices and partial perspectives (Haraway 1988) into contact with each other to produce new knowledges while deepening the current ones. In this way, these gentle gestures and alternative doings may disrupt a forceful drive into the future that leaves a pile of wreckage in its wake.

The first chapter opens with the inception of Singapore's Garden City movement that began with the city-state's independence in the 1960s. It explores how interwoven themes of national development, sustainability, technology, and control (and its limitations) take form in and through urban natures in Singapore, with a particular emphasis on the figure of the garden. The chapter proposes that Singapore's centering on managed forms of nature produces an *authoritarian nature* that flattens the urban environment and more-than-human possibilities. My proposition of authoritarian nature examines the way in which measures of dominance over human and more-than-human communities are bound up with each other. How has Singapore's development and its recent pursuit of world-class status as a sustainable, livable, and eco-modernized city been, in part, enabled by the control of/through natures? In the face of climate change uncertainty, the Gardens by the Bay development, with its giant "Super-trees" and domed artificial environments, becomes a manifestation of control, representing an illusory capacity to persist without serious engagement with the current emergencies of the Anthropocene. At the heart of this chapter, I advocate decentering and undermining singular narratives and forms of control.

Chapter 2 traverses the past, present, and future of Singapore, examining the demolishing of the Kampongs as part of an island-wide public housing development and heritage program aimed at urban renewal. In an ultramodern country that constantly seeks to transform itself to remain at the global frontier, the practices of conservation and the preservation of the natural and built environment, based on its perceived heritage value or bio-value, become highly malleable so as to complement the state's agenda. At the same time, temporality is reworked to fit with a particular modern narrative. This chapter highlights the often-normalized processes of displacement, uprooting, and relocating of humans and nonhumans and their generational, social, and environmental impacts. In particular, it asks how a reductive way of futuring and inhabiting time impacts and disrupts other forms of life. Here, I illustrate the process of *double erasure* that reveals first the literal removal of the built and natural environment, along with the subsequent erasures of culture, memory, and ways of life rooted in these places. These entwined erasures culminate in a kind of amnesia that sustains the narrative of eco-modernization. Through my participation in a program run by a Singaporean social enterprise, I examine how diverse micro-narratives may help to undermine a forced synchronization. What might it take to interrupt the momentum of future (world)-making so as to allow and learn to attend to other ways of life?

Positioned at the intersection of a nature reserve, underground railway, and eco-bridge, chapter 3 focuses on entangled human and other-than-human movements in a condensed urban environment. In 2013, the

Singapore government announced a plan to build the Cross Island Line (CRL), the country's eighth Mass Rapid Transit train line. Since its release, the proposal has caused ongoing heated debate as it involves tunneling underneath Singapore's largest remaining reserve: the Central Catchment Nature Reserve. Intriguingly, as the CRL project threatens to further fragment the reserve, other, more visible, repair work is taking place at the edge of the same reserve: an ecological bridge spanning a six-lane highway that aims to restore animal mobility to an adjoining reserve. Through these two contrasting yet intimately related case studies, the seemingly conflicting acts of connecting and disconnecting enacted in the same reserve demonstrate the complexity and ambivalence of urban mobility, and its entanglement with technology, urban nature, and development. Amid the growing expansion of infrastructure and public transportation in Singapore and around the world, often in the name of sustainability and livability, this chapter unsettles a taken-for-granted, velocity-charged, and human-centered approach to urban movement that is underpinned by the pursuit of a measurable, singular imagination of the future. It asks what it might mean to imagine toward more inclusive and flourishing *multispecies movements* in an increasingly urbanized environment.

As it lacks its own natural water sources, Singapore considers water its most important existential issue. At the same time, Singapore's water story, cited as an exemplar of turning vulnerability into strength through engineering, technological advancement, and central planning, has been widely acclaimed and documented. In chapter 4, I pursue another line of inquiry focusing on the country's development of energy and capital-intensive technological water production to meet its present and (projected) future water demand, as well as the social, environmental, and cultural impacts of this work. Amid escalating climate change effects, including global water scarcity and sea-level rise, technological water—including desalinated water and purified recycled water—is increasingly positioned by the state as a savior, sheltering it from geopolitical uncertainty and untethering it from meteorology. As Singapore cements its identity as a water-maker, its way of engaging with water signifies a new type of climate-resistant water futurity that the island seeks to conjure and export to other regions. This chapter is concerned with the emergence of the *decoupled water* that aims to craft an assumed, illusory ecological invulnerability built on the ethos of separation from nature. How might these types of intensification exacerbate environmental issues? I turn to some alternative ways of relating to water that are bubbling up in Singapore and explore how they might circumvent hyper-separation and intervene in the eco-modernist discourse on water provision.

The final chapter looks directly into the future. In 2017, the government of Singapore unveiled the Farm Transformation Map, a highly technology-driven initiative that is intended to change its current near total dependence on imported food. The plan focuses on the prospect of high-productivity farming, in particular, developing integrated vertical, indoor, and intensive urban farming as possible solutions to the uncertainties of geopolitics, intense urbanization, and environment degradation. Following the stories of a few small- to medium-scale urban farms including rooftop gardens, community farms, and organic farms, the chapter explores the notion of localism and the diverse forms of care being deployed in urban farming in Singapore. As the government's singular planned future for farming seeks to cast out "traditional" farming, deemed as inefficient, this chapter asks: What else might be lost in the process? More importantly, it explores how urban agriculture might forge a thick localism grounded in the work of *situated care* as it carries out social missions, experimenting with and subverting the dominant imaginary of industrial farming.

1

AN ECO-MODERNIZER'S GARDEN

I first "climbed" the Supertrees in 2013, only a year after they were opened to the public. To reach them, I walked through Marina Bay Sands, the integrated shopping and entertainment complex that also holds Singapore's first casino, a highly controversial development. The heat and humidity of the city felt overwhelming. As I approached them, I quickly realized that the Supertrees were unlike anything I had seen before; these tree-like structures—eighteen in all—measuring between twenty-five and fifty meters tall, stood solidly on the ground, among and above the lush "natural" gardens below. Their imposing steel trunks were covered with plants installed on panels. The enormous canopies shaped like inverted umbrellas were unmistakably metallic, extending their silver twigs toward the spaceship-shaped sky park of Marina Bay Sands. These hybrid creatures were disconcertingly composed of both living and dead things, organic and inorganic: I was unsettled by encountering something that felt both foreign and intimate.

Named Supertree Grove, this grand-scale vertical garden is part of an iconic eco-urban project, Gardens by the Bay. Built on reclaimed land in Marina Bay, a prime strategic development area of Singapore, this futuristic complex at a total of 101 hectares consists of three waterfront gardens. The prefix "super" refers to more than the size of the trees; this terminology also aims to capture the embedded environmental functions of the "trees." Some can harvest energy through photovoltaic solar panels, and others are integrated with the two large bio-domes they stand among, serving as air exhaust receptacles (Gardens by the Bay).[1] The first of these biome conservatories is the Flower Dome, the world's largest temperature-controlled glasshouse replicating the cool-dry climate of Mediterranean regions with changing displays of flowers and plants. The second conservatory, the Cloud Forest, features a thirty-five-meter-tall mountain planted with lush vegetation that presents a dramatic altitude change from tropical highlands up to two thousand meters above sea level. The National Parks Board (NParks), a statutory board of the Singapore government, states that Gardens by the Bay is situated "at the heart of Singapore's next phase of development as a global city . . . an integral part of Singapore's 'City in a Garden' vision" (2011). Sustainability is claimed to be at the center of the planning and construction of the project.

The reinforced concrete core of the Supertrees is made to visually resemble an organic tree with the vast covering of plants attached to exterior planting panels. Called a living skin, the vertical greenery attached to the Supertrees consists of over 162,900 plants comprising more than two hundred species (NParks). All the plants have been carefully selected based on strict criteria such as soillessness and high visual interest. They were also extensively tested to ensure their ability to survive and bloom efficiently in a regulated environment. Between the Supertrees is an aerial walkway elevated twenty-two meters above the ground. It offers a panoramic view of the gardens and Marina Bay skyline.

At nightfall, the "trees" light up, transforming the vertical garden into an otherworldly terrain. The nightly "Garden Rhapsody" performance is a symphony of lights, music, and plants. Since its opening in 2012, Gardens by the Bay has become an iconic symbol of the Singapore landscape and one of the most popular tourist attractions not only in the country but also globally, when considering that the city-state itself has been ranked the fifth most visited global destination—based on the Global Destinations Cities Index 2019 (Mastercard 2019). Yet, although the state promoted Gardens by the Bay as a people's garden, its development has sparked discomfort and criticism among some researchers and environmental groups. Much of this criticism has centered on its championing of artificial nature. Some researchers call it an imposed and constructed "nature and history" (Leow

2012), while others describe the greenscape as part of Singapore's "growing technocratic approach to the environment" (Barnard 2014, 7–8) and the country's continuation of "the artificiality of much of the greening process over the past 50 years" (Barnard and Heng 2014, 305). These debates on the techno-centric and artificial aspects of Gardens by the Bay are important. However, Haraway's (1985) concept of the cyborg has long blurred, complicated, and problematized the boundary between the human and the non-human, the biotic and the machinic, while also attending to the important work that these categorizations continue to do. It is crucial not to overlook other critical issues and implications relating to this futurist garden project. As discussed in the introduction, there is a growing body of scholarship that is moving away from the debate over whether or not something is natural, and instead addressing the question of which natures are being produced and their uneven impacts on diverse communities (Macnaghten and Urry 1998; Wolch 2002; Haraway 2008; Davison 2008).

In its focus on Singapore—variously presented in official accounts as a "Garden City" or a "City in a Garden"—this chapter explores how the interwoven themes of national development, sustainability, technology, and control (and its limitations) take form in, and through, urban natures in Singapore, with a particular emphasis on the figure of the garden.[2] Inspired by eco-feminist work on mutually reinforcing forms of mastery over nature and humans, I propose that Singapore's emphasis on managed forms of nature produces an *authoritarian nature* that flattens urban environments and their more-than-human possibilities.

The first section offers a brief discussion of Singapore's spatial limitations, obsession with development, and greening practices. It examines how, since the country's independence, the inception of the Garden City movement has given rise to an authoritarian nature. I explore how the imagination of the garden plays a significant role in shaping the nation as an idea and a reality, while also cultivating specific types of citizens. The second section further develops this concept by examining the historically shifting form of authoritarian nature during Singapore's transformation over the past decade into a City in a Garden. This more systematic and technological greening scheme aims to cement Singapore's status as a leading sustainable, livable, and eco-modernized city. Taking a detour, this section also turns attention toward some other possible plant-human relations through the fugitive seeds that traveled across borders (Keeve 2020).[3] The final section shifts the focus back to Gardens by the Bay to explore the ways in which this facility becomes an embodiment of the state's authority and technology in terms of naturing and futuring. In the face of climate change uncertainty, I argue that the domes beneath the Supertrees, and their perceived stable environments, are enabled by authoritarian natures, and in turn help the state to validate the necessity of dominance. As I conclude the chapter, I point to the importance of attending to various nonhumans who resist and escape this control to subvert the modality of authoritarian nature—a theme that is further taken up in later chapters.

THE BIRTH OF AUTHORITARIAN NATURE

Continuing my exploration of the urban gardens and the green future of Singapore, I head toward Singapore City Gallery (SCG). There, the interactive exhibitions offer generous and detailed accounts of the dramatic physical transformation of the young city-state since its independence in the 1960s, from a trading post founded by the British to a global city with an ultra-modern, green, and clean image. Walking from room to room, the exhibition explains that the focus of each stage of the city's development was to optimize the use of land and space and to overcome resource constraints on economic development. Visitors are repeatedly reminded of the paramount importance and efficacy of the People's Action Party (PAP), Singapore's single ruling party since independence and its central planning and strong leadership. In addition to creative land allocation, Singapore has grown in physical land mass by more than 21 percent since the 1960s through aggressive land reclamation from the sea. As a result, any sign of the "natural" shoreline has disappeared.[4]

In Singaporean sociologist Chua Beng Huat's (2011) account, the city-state has transformed the extreme consciousness of its "zero" resources and intense space-scarcity into the urgency for survival, creating the need to stay ahead of time and secure a competitive position. Some suggest that this kind of narrative of vulnerability has created a "siege mentality" that legitimates the authoritarian rule of the PAP (Brown 1998; Neo 2010; Heng 2013). Indeed, Singapore's single-party government possesses absolute power in planning and execution, from the remolding of its landscape to mega public housing schemes enabled by drastic social and physical engineering.[5] It is in this context of scarcity, a belief in control, and a desire for persistent economic development, as well as the need to overcome perceived spatial and resource limits, that Singapore's favoring of a modern, efficient, and orderly form of greenery needs to be understood.

Since independence, Lee Kuan Yew, the first prime minister of the young city-nation, was determined to create a clean, green, and orderly city. In 1963, surrounded by a large crowd (probably his officials), Lee planted a Mempat tree, initiating a national greening movement and the Garden City Scheme (Singapore Press Holdings).[6] As Belinda Yuen points out, the garden is favored as it is "seen to be a form of greenery compatible with urban living, one which has no 'unruliness' and is shaped and controlled by human hands" (1996, 962). After planting millions of trees and building hundreds of parks over the past decades, Singapore is widely recognized as one of the greenest cities in the world, providing the conducive and stable environment needed for its economic growth.[7]

Some researchers contend that Singaporean leaders "have manipulated the natural environment to reflect their vision of a planned green, urban landscape" (Barnard and Heng 2014, 283). Others have referred to nature in Singapore as completely commodified through "the production, distribution, and exchange of biophysical green resources" (Gulsrud and Ooi 2015, 88). In this dense city-state, any tension between conservation, maximizing urban greenery, and continuous development has been further exacerbated by the different perceptions of nature held by the Singapore government, citizens, and conservationists (Neo 2007).[8] As Aidan Davison and Ben Ridder remark succinctly, "There is, in effect, no single urban nature. There is an array of contested urban natures" (2006, 307). With a dominant imaginary of Singapore as green and livable, it is important to complicate the discussion of urban natures so that they can't be so readily flattened and rolled into dominant narratives and strategies.

In this chapter, I further these debates and propose the concept of *authoritarian nature*. Environmental philosopher and ecofeminist Plumwood's work on the inextricable relations between various forms of domination/oppression is central to my thinking: "The way that control over and exploitation of nature contributes to, or is even more strongly linked to, control over and exploitation of human beings" (1993, 13).[9] Other feminist environmental researchers have highlighted the way in which not only women but various other groups of people are oppressed within these systems, which has become a crucial premise in approaching the issues of water, food, and environmental injustice (Cuomo 1998; Gaard 2001; Green 2020). Scholars from other disciplines have also explored this topic and provided powerful insights. For example, political ecologist Paul Robbins explores the entangled domination/civilizing of nature and people and reveals how suburban lawns create "rather than simply catering to, a certain kind of citizen" (2007, 25).

In particular, exploring authoritarian nature through the trope of the garden, the following sections examine how modes of domination become entangled with each other and how authoritarian nature directs a material-semiotic imagination of urban nature that has been fundamental to Singapore's nation building and development. In Singapore's pursuit of economic success, nature is to a very significant extent directed from the top down. In such a context, the question is which natures will be valued and kept, and how will they be kept and where? Moreover, how do these particular forms of authoritarian nature shift and change over time?

As a middle space, a liminal bridge uniting the divided terrain between human and nature, the garden has been a cultivated, measured spatial and architectural form that has played a significant and intimate role in the social-cultural imagination since ancient times (Foucault 1986; Hartigan 2015, 2017). In particular, the garden has often served as a powerful and effective way to discipline people and produce good citizens and preferred forms of society (Stoetzer 2022). In the late nineteenth century, Ebenezer Howard's garden city movement, which advocated building a series of cities surrounded by green belts, was proposed as a solution to the overcrowded and heavily polluted British cities. To Howard, the "joyous union" of human society and the beauty of nature "will spring a new hope, a new life, a new civilization" (1902/1965, 48). Political scientist James C. Scott's (1998) illuminating work on authoritarian planning of utopian projects draws an analogy between social engineering and gardens.[10] Zygmunt Bauman (1989), a notable thinker on modernity and its violence uses the metaphor of the "gardening" state to describe the modernist desire for order-making, practices of planning, dominance over society, and the effort to weed out the undesirable.

For contemporary Singapore, the envisioning of a modern garden city since the 1960s has been central to its socioeconomic development, motivated by a desire to attract foreign investment and highly skilled labor. In Lee's words, "one arm of my strategy was to make Singapore into an oasis in Southeast Asia, for if we had First World standards then businesspeople and tourists would make us a base for their business and tours of the region" (2000, 173–174). The Garden City movement also served as a tool for land use allocation. Over the next two decades, a large amount of forest, swamp, and villages were cleared to make way for built-up areas including industry, transportation, public housing, and commercial zones in the Singapore mainland (Yuen 1996, 956). The process of urban greening in the mainland was also enabled by the relocation of various heavy industrials and waste processing centers along with spaces of leisure to offshore islands that were primarily "inhabited by vibrant Malay and *Orang Asli* (indigenous people) communities" (Connolly and Muzaini, 2022, 6, italics in original). This is a point I will return to in chapter 2. Meanwhile, other forms of nature were "built" to green the city, including parks, extensive planting of roadside trees, and the planting of greenery to camouflage concrete structures such as overhead footbridges (Yuen, 956). In the following half century, continuous expansion and transformation of the garden city formed an essential part of Singapore's story of transition from the third to the first world.

As David Harvey notes, "There is always an authoritarian edge somewhere in ecological politics" (1993, 21). While this may be the case, Lee's imagining of the young city-state as a garden city, and himself as the chief gardener, takes this alignment to a new scale and a specific intensity that marks the birth of authoritarian nature.[11] The ambition of Singapore's garden city movement was not only to reshape the landscape but also, as Lee himself put it, to alter the "rough and ready ways of the people" (2000, 174). To complement the Garden City movement, national campaigns such as "Keep Singapore Clean" were powerfully mobilized. A system of financial incentives and disincentives accompanied the campaign's "media blitz" to "persuade the people to adopt the attitude and behaviour recommended by the campaign" (Lim 2013, 5). Here, the figure of the garden not only provides a pleasant façade but also enables "a moral urban politics based on the enrolment of subjects into 'civilized' behaviour" (McFarlane and Rutherford 2008, 367). Furthermore, building an orderly and clean urban environment helps to justify the authoritarian approaches toward natures and the citizens in important ways. For example, this included demolishing the Kampongs (villages in Malay), described as unruly and unhygienic, and relocating farms and backyard trade industries along the Singapore River, whose way of life was identified as a source of pollution (Loh 2009; Tan, Lee, and Tan 2016).

As we learned from Lee, one reason that the garden was chosen as the tool for cultivating the type of citizen and nation he conceived of lay in the level of care and attention required to maintain the garden: "The grass has got to be mown every other day, the trees have to be tended, the flowers in the gardens have to be looked after so they know this place gives attention to detail" (cited in Kwang, Fernandez, and Tan 2015, 12). After decades of greening and developing, authoritarian nature has seen the country rise to be one of the greenest in the world. But it has also helped to cultivate the Singaporeans as modern citizens with the "nationally inflected personal qualities of being hardworking, efficient, and effective" (Chua 2011, 32). In controlling nature and the citizens in this regimented and interrelated way, Singapore effectively creates the sense of a highly secure, stable environment in which to invest, live, and visit. In this way, controlling nature gives a reassurance to the outside world and to locals.

One particularly stark example of this control can be found in the curation of the city's trees. As in many parts of the world, the state decides on the exact type of trees to plant, where to plant, and for what purposes. But in Singapore it also controls the temporal span of the plants: the exact moment of their lives that are required for the city. Writer Timothy Auger summarizes this succinctly, "Trees take time to grow and Singapore needed quick results" (2013, 33). These two deeply conflicting temporal needs have seen tree removal and transplanting become common practices in Singapore, from relocating gigantic rain trees (*Samanea saman*) for the rejuvenation of the historical civic district, to importing "instant trees" as a way to compress time (Auger; see also Yuen 1996).[12] Some 70 percent of trees already planted in Singapore today spent their early years outside of the city-state, mostly planted out in tree banks in Malaysia on vacant land until they had grown to a sufficient size to meet the diverse needs of the upcoming parks, residential, and industrial developments (Auger).

To create, cultivate, and maintain a neat and orderly garden, importation does not stop with plants. Singapore's economic success and its transformation into a knowledge-production model in the 1990s resulted in the planting and maintenance of the gardens being largely outsourced to foreign laborers, as are many other aspects of the city from construction to diverse types of care work. As Miriam Williams succinctly puts it, "Who does the work of care in our worlds" is deeply political and classed (2020, 1). Although the garden environment is specifically maintained to attract foreign investment and the associated immigrants, referred to as "foreign talent," Singapore adopts distinct policies governing the lives of "foreign workers," who are considered low-skilled (Yang, Yang, and Zhan 2017). While foreign talent is positioned as a contributor to the nation's development and encouraged to integrate as part of the population mix, the foreign worker is "structurally built into the economy as a transient category performing work shunned by, and at a wage unacceptable to, the local labour force" (Yeoh 2006, 32). Akin to the imported trees whose lives are closely controlled, the foreign workers have been "prevented from putting down any roots in Singapore society" as they are "barred from marrying a Singaporean or Permanent Resident, or becoming pregnant" (Yeoh and Soco 2014, 173).[13] In a vein comparable to the approach toward the composition of the garden city, who is allowed to take root in the city-state, when, and for how long, are tightly managed affairs, situated in the larger context of social inequalities and exploited human and other-than-human labor in Singapore.

Furthermore, seeing the urban through a more-than-human lens requires not stopping at humans' attempted mastery over, but also through, natures. In Plumwood's account, nature should be integrated "as a fourth category of analysis into the framework of an extended feminist theory which employs a race, class and gender analysis" (1993, 1–2). In the context of urban greening in contemporary Singapore, control from the state is not simply about natures and/or citizens. The foreign labor, employed to help produce the garden and then subjected to it through the ongoing process of maintaining the greenery and the city, are also deeply implicated. Indeed, imported trees, transplanted trees, and the transit foreign workers have all become a kind of "mobile assets," dominated through instrumentalism and incorporation (Tsing 2015, 5; see Plumwood 1993). Crucially, attending to entangled forms of domination requires going beyond the singular human. In Singapore, foreign workers and diverse local citizens are not dominated in the same ways. Rather as Plumwood insisted, this is not about human/nature dualism in any straightforward way but about multiple, distinct, and mutually reinforcing forms of oppression.[14]

Equally, it is to attend to the agency of the urban greenery and, to borrow environmental anthropologist Sophie Chao's words, (re)think the (capitalist) dominance "as a multispecies act" (2022, 6).[15] During the anthropologist Natasha Myers's visit to the Gardens by the Bay, she noticed the foreign workers hanging from the indoor human-made mountain to maintain the plants and decried that the green living infrastructure "itself thrives on the energetic, material, and affective labours of marginalized people" (2015, 34). Here, within the capitalist and (eco)modernized greening scheme, not only are some urban natures controlled but their flourishing also relies on their unequally distributed dominance over different groups of humans.

As the creation, experimentation, construction, and maintenance of the city garden is largely outsourced to foreign labor, trained experts, and increasingly technology, the need for care and attentiveness associated with gardening diminishes the encounters between certain groups of humans and plants. This mode of greening not only disentangles the relations and erases the "intimacy between humans, land and creatures" that gardening may provide (Ginn 2017, 4) but also introduces further modes of controls.

Unlike Howard's garden city vision that focused on creating a society of "good, righteous, and healthy citizens" through urban spatial planning rather than the details of the garden itself (Christensen and Heise 2017, 437), or the use of the garden as a metaphor for social structure in other projects, the imagining of Singapore as a garden city has made the garden and modes of naturing an integrated part of the ongoing nation-building project and mastery over both human and nonhuman. Through forms of oppression and hierarchized dualism, authoritarian nature has been built into "a certain kind of logical structure" in the nation (Plumwood 1993, 41). Singapore presents a site that invites us to rethink how control of diverse humans is enabled through the figuration of the garden while natures are simultaneously subjected to mastery.

GREENING WHILE GROWING

As part of the program of the International Green Building Conference 2017, a large high-profile urban sustainable event (see introduction), participants were able to visit buildings that had been awarded a Green Mark Platinum for their environmental features and performance. This is the highest accolade for green building certification sanctioned by the Singapore government (Building and Construction Authority).[16] Along with other attendees including policymakers, business delegates, and researchers, I visited the CapitaGreen skyscraper, located in the heart of Singapore's central business district on the site of a former car park. Like some other high-rise buildings endowed with lush greenery in Singapore, CapitaGreen looks like a vertical forest from afar. Along with multiple sky parks and extensive indoor greenery, large plants have been planted between the double skin facade that is designed to reduce heat absorption while also creating something of the feeling of working in an open forest for its tenants. On the rooftop, a magnificent giant funnel sculpted like a blooming flower converts the dynamic pressure of the wind into positive static pressure, then channels fresh cool natural air into the office floors for optimal indoor air quality and a conducive working environment.[17] The roof also holds an urban forest. A chic restaurant nestled among the trees offers a "sky forest" dining experience complemented by panoramic city views.

Not only is CapitaGreen replete with greenery but it was, according to the architect, conceived as a visual metaphor for a tree. On the ground level, a commissioned artwork by a renowned Italian artist consists of clusters of slender, tall sinuous poles representing roots and shoots. The earthy-colored walls of the building, made by master craftsmen from Japan, specifically use Singapore's own soil. In the account of a senior staff member of CapitaGreen, this bio-hybrid skyscraper seeded in the island's own soil endowed with lush vertical greenery and plants is an exemplar of the city's vision of living in harmony with nature. Since its opening, the ecological features of CapitaGreen have attracted some highly influential companies as tenants, for instance, Lilian Tham, the chief operating officer of one tenant, Schroders Investment Management, expressed the company's motive for leasing the building clearly: "The building's environmentally-sustainable features appeal to our values as a responsible corporate citizen" (CapitaLand 2015).[18]

In the face of challenges from other mega cities and mounting ecological concerns arising from intense urbanization, the city-nation advanced from a "Garden city" to a "City in a Garden." With the importance of urban natures increasingly acknowledged in building a sustainable city, the highly suggestive conceptual shift from a city-centered to garden-centered space has shown the world Singapore's determination and ambition to define a sustainable urban green narrative and to maintain its position as a world city. In the World Cities Summit 2012 held in Singapore, Singapore's prime minister introduced the "City in a Garden" move as aiming "to bring green spaces and biodiversity to our doorsteps." Although some have called the transition introduced around 2008 a mere symbolic movement (Chua 2014), the proposition in fact translates into a comprehensive, multifaceted, and techno-centric naturing scheme driven by a continuous desire to open up urban space for further development while simultaneously attempting to craft a particular eco-modernist future.

According to the NParks, the City in a Garden scheme includes continuing planting trees and building more parks, focusing on building world-class gardens, in particular, the development of the futurist Gardens by the Bay (see figure 1.1).[19] The scheme also involves building a larger network of Park Connectors and Nature Ways in response to the city's highly fragmented green patches due to its legacy and ongoing development.[20] Among the many measures, a prominent aspect of the City in a Garden is to create a "'Vertical Garden City' through active greening of vertical space" including green walls, green roofs, and rooftop gardens (Tan, Wang, and Sia 2013, 29). Singapore's upward growth to counter its high density embraces the development of skyrise greenery technology. To ensure a wide adoption of skyrise greenery, a range of incentive schemes from various government agencies have been introduced to encourage building owners to add rooftop greenery and vertical greenery (Cossé 2011). The aim of being a city in a garden also includes carefully guarding the number of plants through a Landscape Replacement Policy in certain areas, meaning that some ground greenery demolished as a result of development in selected areas needs to be replaced by other forms of greenery, which can include skyrise greenery (Urban Redevelopment Authority 2017).[21] As with

1.1   At nightfall, the Supertrees light up. *Source*: photo by author.

other aspects of Singapore, "the discussion of sustainability is tightly tied with the usage of land and space" (SCG).[22] Hence a central inspiration and motivation in developing skyrise greenery technology is to increase useable space in a dense high-rise environment.

Some Singaporean planners have championed vertical green technology as a novel ecosystem that helps to safeguard future biodiversity, offering a "potential model" for high-density cities that face the challenges of green space (Hamnett and Yuen 2019, 20). In the words of Peter Newman, an urban planning researcher, the wide adoption of skyrise technology in Singapore suggests "a city where a new kind of urban nature develops, which fulfils the functions of the original ecosystem replaced by the city and contributes to local biodiversity improvements" (2014, 64). However, in a study of diurnal birds and butterflies encountered in roof gardens of Singapore, ecologists caution that conservation goals for green roofs need to "take into account that these novel ecosystems may be environmentally predisposed to host certain suites of species, rather than aiming to directly replicate ground-level ecological communities" (Wang et al. 2017,

17). A local conservationist who does not object to the concept of vertical greening pointed out to me that the replacement of ground-level forest by the same amount of roof garden does not support the ecosystem in the same way.

In their studies on urban biodiversity, Nicholas Williams and colleagues (2014) observe that green roofs in various parts of the world have greater species diversity than conventional roofs. However, whether they may support similar biodiversity as ground-level habitats, and their potential to connect ground-level habitats, is largely untested and not clear (Williams, Lundholm, and MacIvor). In this light, Williams et al. urge that "green roof proponents should use restraint in claiming conservation benefits and it is premature for policymakers to consider green roofs equivalent to ground-level urban habitats" (1643). Recent research in Melbourne and elsewhere further suggest that the size, location, height, and maturity of skyrise greenery and their interconnection with surrounding habitats are all important factors in their effectiveness in supporting biodiversity conversation (Dromgold et al. 2020).[23] It becomes clear that more comprehensive and ongoing examination of the benefits of skyrise greenery is needed. Here, what is concerning is how the development projects, through demolishing and replanting, already perform the kind of "homogenisation" work that Plumwood warns against (1993). That is, different forms of greenery and their associated ecosystems are erased and defined "*in relation* to humans" (Plumwood, 70, italics in original) as green technology required for continuing urban development.

In a similar vein, in 2017 it was reported that roughly 10,000 to 13,000 trees in Singapore could be removed over the next fifteen years to make way for various development projects (Toh 2017). Nevertheless, the loss of these trees will not be taken seriously as the NParks promises that all affected trees will be replaced on a minimum of one-for-one basis (Toh). Akin to the idea of interchangeable ecosystems that in part guides the implementation of skyrise greenery, the reduction of plants and forests is carefully framed in a narrow, mathematical sense that simultaneously emphasizes and diminishes urban natures. This kind of control in greening calls to mind the controversial yet increasingly popular scheme of biodiversity offsetting that allows development work if some form of conservation activities "are undertaken elsewhere to compensate for the environmental impacts" of the development projects (Benabou 2014, 103; Apostolopoulou 2020). In the context of Carnaby's Black cockatoos in Perth, Australia, Donna Houston (2019) argues that this kind of scheme, regrowing bushland in another area to compensate for those cleared for urban development, ignores the temporal rhythm of the plants, which may ultimately lead to the disappearance of the Black cockatoos that rely on them. Thus, Houston appeals for rethinking in urban planning given the shadow of extinction. Although biodiversity offsetting is a different aspect of conservation, planning, and environmental justice, a familiar reductionist view toward a more-than-human relation at work in Singapore is alarming.

Indeed, urban skyrise greenery, though a relatively new mode of greening, may very well offer various benefits including reducing urban heat island effects and improving rainwater retention (Newman 2014). What I take issue with is not vertical greenery (or tree planting) but rather how it is used as a new form of dominance that allows the freeing up of space at ground level to enable growth or development. This assumes that the existing ecological systems can be easily replaced in another form at another location. At the same time, green buildings such as CapitaGreen are seen as exemplary by bringing nature closer while growing the economy, providing residents with a semblance of living with nature while satisfying their obligation to be responsible citizens or companies. In short, skyrise greenery encapsulates the ethos of an eco-modernist and anthropocentric way of engaging with nature. It is crucial to situate the push for techno green systems within the ongoing issues of the reduction of mature trees in Singapore (Tan, Wang and Sia 2013) and the country's relentless desire for continuous progress.

Furthermore, part of the imagining of living in a garden is envisioned as being in harmony with nature including increased biodiversity. To this end, Singapore pioneered the City Biodiversity Index, a self-assessment tool to monitor urban biodiversity and to guide conservation works (NParks 2010).[24] As there are only a small number of natural reserves containing Singapore's flora and fauna that are legally preserved from development, much emphasis has been placed on streetscape planning. The staff member in NParks explained to me that meticulous planning work had been done on tree provisioning. For instance, ensuring that the right type of trees are planted to attract the "right" animals to enhance biodiversity.[25]

As cultural geographer Franklin Ginn points out, "gardening means chan-
nelling one possible future over another; a garden excludes certain beings,
denies others the chance for life, even as it extends hospitality to some"
(2017, 2). As I have discussed in the introduction, there are continuous
efforts to remove those deemed unwelcome in the garden. In this light,
"homogenisation" is only part of a much larger scheme to shift authoritarian nature into focusing on crafting and creating an eco-modernist garden
and shaping a particular imagining of urban sustainability through intense
controlling of/through nature.

The vision of a City in a Garden continues to bring to the fore the tension between conservation and continuous development as many of its
purported sustainable practices are accompanied by ongoing loss of certain types of greenery and the expansion of infrastructure with uncertain
sustainability features. For example, part of the Gardens by the Bay now
occupies what was once marsh and wetlands that provided critical habitat
for a host of bird, butterfly, and dragonfly species, some of which were
rare and endangered (NSS 2010). Many years prior to the current development, it was also the site of the Marina South duck ponds, an eleven-
hectare piece of land-turned-roosting-ground for rare migratory ducks that
was later reclaimed and paved over (Gulsrud and Ooi 2015). In response
to an inquiry from a local environmental group, the ministry of environment argued that as the ponds were originally dug out by humans, there
was nothing natural about the site and therefore the site did not qualify
for conservation status (Neo 2007, 195).[26] Political ecologists Natalie Marie
Gulsrud and Can-Seng Ooi maintain that prioritizing concrete Supertrees
over wetlands symbolizes the tension within Singapore between protecting sensitive ecosystems and wildlife habitats and commodifying nature
for recreational opportunities and tourism (2015, 84). To me, it is important not to approach the seemingly different ways of treating the Gardens
by the Bay, the marshland, and the pond as a mere tension between wild
pristine nature, an imagined state that itself has been commercialized and
commodified, and an artificial nature. Rather, as Davison and Ridder state,
"Urban nature itself plays an important role in the creation and evolution"
of human ideas and desires (2006, 307).[27] The strategy of the state, in this
context, shows that there is both a desire for and devaluing of "artificial"
nature, as long as it is convenient and/or profitable.

With the increasing attention given to environmental issues, the envisioning of a City in a Garden captures and co-opts the desire to enact a sustainable future while remaining competitive without compromising development. In the process, it effectively influences people's perceptions, so they gradually accept the kind of urban nature presented to them. This in turn contributes to furthering the authoritarian approach toward urban sustainability. The trajectory from Garden City to a City in a Garden, entwined with the dynamics of social, economic, technological, and environmental change, demonstrates how the state has thought to flatten a mode of naturing and distorts the composition of the garden. In the process, the mutually reinforcing mastery of more-than-human communities is normalized. If the Garden City movement reflects a type of authoritarian nature that prioritizes neat and orderly forms to discipline the residents and establish a stable environment for socioeconomic development, the City in a Garden underpinned by new technologies (like vertical greenery) enables dominance over urban nature through an assemblage of homogenization, targeted diversity, and meticulous work of inclusion and exclusion as it seeks to enact an eco-modernized sustainable green future.

Despite these comprehensive measures, control is never absolute. Before moving to the next section of the Gardens by the Bay, I would like to take you to another patch of green momentarily—a proclamation garden—to see how plants, borrowing urban anthropologist Bettina Stoetzer's words, "are both followers and fugitives of culture" (2022, 70). Indeed, some plant residents of "the city in a garden" are creating their own paths of rooting and becoming.

In 2019, Singaporean artist Charles Lim Yi Yong exhibited the latest piece of his *SEA STATE* series—SEA STATE 9: proclamation Garden—at the National Gallery of Singapore.[28] Since 2005, Charles Lim's *SEA STATE* series, consisting of photographs, maps, films, sculptures, an archival website, and more have been exploring the complex and negotiated "biophysical, aspirational and cerebral contours" of Singapore in the context of the city-state's radical urbanization and development.[29] Continuing and expanding his meditation on space, border, and sea, the new living art installation turns attention to the contentious issues surrounding land reclamation in Singapore through the lens of plants and plant-human relations.

To "grow" the proclamation garden, the artist, working with local bota-nists, curators, and landscaping companies, replanted some less-known species of plants that were found in reclaimed areas across Singapore to the roof garden of the gallery, transforming an often orderly space to a messy assemblage of growing plants and human visitors.[30] Among these vegeta-tion, some species were previously thought to be extinct in Singapore; others are rarely seen, such as date palm (*Phoenix dactylifera*) (Tan 2019). According to Adele Tan, the curator of the exhibition, these plants, often considered as "exotics," were probably brought into Singapore "acciden-tally or intentionally" by humans (2021, 123). For example, the seeds of these plants may have come in with foreign sand that was subsequently used in land reclamation; or they may have come through the food that migrant construction workers "who toil on the land habitually eat," and the plants may have then germinated after workers "spat out the seeds" (Tan 2019, 2021, 123).

While it might be thought (or expected) that the "new" reclaimed area is an empty space waiting to be developed by humans, the reality on the ground can be quite different. As local botanists, geologists, and environ-mental historians note, given that reclaimed land was often required to stabilize for years before being used for development work, many plants, despite the harsh environment, started to grow and some even became sizeable forests, making home to many other beings in the process (Yee et al. 2016; Gaw and Richards 2021; Powell 2019). Taking root in the bor-rowed land from the sea, these seeds, which traveled from afar or were dispersed locally, are not subjected to be governed. Rather, their germina-tion and multiplication seem to embody an urban "ruderal ecology"—a form of nature emerged in disturbed environments—that Stoetzer sug-gests as capable of disrupting "systems of human control" and hegemonic human-environment relations (2022, 4).[31]

Although the plants of the proclamation garden had to be removed from the National Gallery at the end of the exhibition, their ruderal pres-ence signals and/or makes visible other kinds of human-plant entangle-ment in Singapore. At various reclaimed sites across the city-state, plants, migrant workers, imported sand, and the sea continue to co-shape these new, anthropogenic urban landscapes, countering (even for a little while) the state's authoritarian practices over the movements and lives of many.

THE (DIS)APPEARING CLOUD FOREST

Among all the eco-urban developments, Gardens by the Bay is one of the most visible on the global stage. In the official narrative, these hybrid gardens capture "the essence of Singapore as the premier tropical Garden City with the perfect environment in which to live and work—making Singapore a leading global city of the 21st century" (Gardens by the Bay). In the BBC documentary *Planet Earth II* (2016), accompanied by a 360-degree view of the Supertrees and their surroundings, David Attenborough narrates rhythmically, "Perhaps the most spectacular example of city greening is this growth of Supertrees.... This is a new urban world that we have now designed and built with others in mind." As viewers enjoy a lavish panoramic shot of Gardens by the Bay, Attenborough continues, "Is this a vision of our cities of the future? It could be possible to see wildlife thriving within our cities across the planet." Fredi Devas, the producer of the *Planet Earth II* program, spoke of his wonder at Gardens by the Bay to the local media in this way: "What really attracted me to Singapore was the ethos of building a city within a garden—the idea that Gardens By The Bay, for example, has been built before the skyscrapers that are going to be built around it . . . I think we're really championing city greening in a futuristic way."[32]

Intriguingly, the two bio-domes of the futuristic garden contain tens of thousands of plants that would not normally exist in Singapore's tropical climate. Many of these plants are endangered in their own habitat due to deforestation and environmental destruction. Prior to the launch of Gardens by the Bay, six prototype glasshouses were used for extensive testing of advanced cooling technologies as well as optimum temperatures and humidity for the conservatories (Transsolar 2013). The first shipment of plants for the Gardens arrived in 2008: a $2 million collection of bromeliads native to the Americas, consisting of 210,000 plants of more than three thousand varieties (Chua and Yap n.d.). At the opening of Gardens by the Bay, Tan Wee Kiat, the chief executive of the eco-project, says, "Here on the equator . . . we're in a garden that is perpetual summer. Into this garden, we've brought two glass houses that give you a touch of perpetual spring" (cited in Lim, 2012).

Despite the intention of immersing the locals and visitors in a "perpetual spring," the two bio-domes are also said to tell "stories" of human, plant and planet, and the threats and consequences of climate change (Er et al. 2013). For instance, the futuristic Cloud Forest, one of the two conservatories simulates and pays tribute to a unique habitat—cloud forest—that is fast disappearing due to climate change and deforestation (Gardens by the Bay).[33] It also aims to "generate awareness of the dire state of cloud forests around the world" and the effects of climate change including biodiversity loss and extinction (Gardens by the Bay). In reality, a visit to the bio-dome offers a quite different experience.[34] As soon as visitors enter the dome, they are welcomed by an awe-inspiring sight: the world's tallest indoor waterfall shrouded in mist and a thirty-five-meter-high mountain covered with a sensational number of plants. The cool air in the dome immediately gives a much-needed respite from the humid local climate. After visitors take photos with the misty mountain as background, they follow the ascending path circling around the mountain admiring divergent rare and foreign or endangered plants growing in an unlikely habitat in a controlled atmosphere. Just before visitors are about to depart the dome, they pass an exhibition room with educational audio and videos on the effects of global warming and disappearing plants. However, eclipsed by the technologically reliant botanical display or shadowed by the light and music show at night, the messages of environmental destruction and extinction are as precarious as the state of these issues themselves, backgrounded and barely visible.

Moreover, as the conservation dome totally controls the environment, from temperature to airflow, humidity, and plant species, it not only offers the visitor a respite for tropical heat but also provides a seeming relief from the effects of climate change. In this context, the bio-dome performs what Myers describes as a magic trick, distracting the visitors from the slow violence "on the other side of its gleaming glass exterior" (2019, 133; see also Lim 2014).[35] In other words, it creates an illusion for the populace and visitors from other parts of world that humans have the capacity to conjure any kind of nature they might desire, or to sustain a kind of nature that is fast disappearing outside the dome. Although it could have opted for another type of high-tech tourist attraction, the Singapore government chose this plant- and nature-centric project to emphasize the success of its decades of greening practices and its ability and determination in continuing this mode of naturing. Built on reclaimed land, Gardens by the

Bay along with the bio-domes have, in part, been reworked as a powerful embodiment of the state's authority over ecosystems using technology and wealth.[36]

The conception and construction of a certain technology-assisted natural environment for the amenity of humans brings to mind another iconic dome. Following the end of the Cold War, in the midst of a new round of power struggles involving space exploration and the enthusiastic pursuit (or preparation) for humans colonizing extraterrestrial space, Biosphere 2, the largest enclosed artificial ecological system to date was launched in the Arizona desert (Poynter 2006). The goal of the project was "to create an ecosystem and human habitat as good as, or superior to nature on Earth (known in this context as Biosphere 1)," in the hope of developing "'a prototype for a space colony' and to settle future 'sustainable communities' on Mars" (Höhler 2010, 39 and 46; see also Allen 1991). It existed for just over a decade, during which time two human enclosure experiments were carried out. Throughout the project, the composition of nature was systematically calculated and finely tuned to ensure the survival of humans. What to include or exclude was determined by, firstly, whether the element was essential for the self-sustaining ecosystem, and secondly, whether it needed to be regenerated "naturally" inside the dome. According to science and technology studies scholar Sabine Höhler, polar regions, for example, "were not modelled," as their primary purpose "on earth was considered to be climate regulation," and this was "substituted by technology" (2015, 121).

Although created decades apart, the similarities between the two artificial bio-domes are striking. Both ecological systems spring from modernist imaginaries focusing on pushing humans' ability to create a controllable nature in which only selected elements are included, thereby ensuring the desired balance and stability. Neither project was conducted with any real intention of engagement with the external environment. Rather, they function as a kind of preparatory project that selectively re-creates and establishes something in a different region that is either unavailable or lost, be it earth or a cloud forest.

My interest, however, is less in what these domes have in common and more in examining the differences between them. While the substantial cost of building and the subsequent ongoing maintenance of Biosphere 2 contributed to the termination of the project, the Singapore government is well-versed in the need to pair capitalism with the construction and maintenance of authoritarian nature. Even though the cost of Bay South Garden alone is estimated to be a billion Singapore dollars (US$829 million) and attracts a hefty ongoing maintenance fee, most of the costs are being recuperated through tourism. Indeed, Gardens by the Bay has even been made into a valuable generator of profit. The entrance rate for both biomes is S$28 for nonresidents and an additional S$8 for a quick stroll on the OCBC skyway. Listed consistently by TripAdvisor—the world's largest travel information site—as Singapore's top attraction, Gardens by the Bay has welcomed more than fifty million visitors to its inspiring futurist vision since its opening. This also suggests that this eco-development in fact is dependent on tourism, an industry that itself contributes significantly to the global carbon emissions that are threatening cloud forests and biodiversity more generally around the world (Lenzen et al. 2018).

While Biosphere 2 was criticized as a "glass coffin" and a minimalist techo-lifestyle by Jean Baudrillard (1993, 250–251; see Höhler 2015, 127), Gardens by the Bay is a capital-techno-ecological exercise in city-branding and tourism that, in turn, disguises an ecological concern. In the account of Kenneth Er, the chief operating officer on the project, Michelle Lim, and Andrew Grant, the architect behind the Bay South Garden, the Supertrees demonstrate human's "attempts at recreating nature's balance" (2010, 36). The state calls Gardens by the Bay a people's garden that represents the effort to encourage people to be closer to nature. Here, it is important to consider the kind of natures that are being pushed to the fore. What do they inspire and instigate? Furthermore, it is problematic to have an eco-project that itself is inherently unsustainable, both from the energy consumption perspective and its high reliance on tourism. Although it is said that "a suite of cutting-edge technologies" have been deployed to minimize the building's environmental footprint, in particular in cooling, the energy consumption could only be somewhat reduced in comparison to conventional cooling technologies (Gardens by the Bay).[37] In addition, Gardens by the Bay and the number of plants on display, mostly nonnative, require intense ongoing efforts to maintain. Seen in this light, the eco-message Gardens by the Bay seeks to sustain is ambiguous at best.

Where the failure of the first experiment of Biosphere 2 was partly due to participants having to rely on oxygen being injected from external sources, Gardens by the Bay and Singapore have managed to create a powerful image of a self-perpetuating system, portraying a dangerous imagined capacity to persist without the need to seriously engage with the current emergencies of the Anthropocene. For example, it neither fosters a thickened relationship with the environment nor addresses the country's culture of high consumerism. Alarmingly, a recent study mapping the changes in global deforestation footprints covering the period of 2001 to 2015 highlights the consumption patterns of Singapore, along with other wealthy countries, as a significant and growing threat to tropical forests (Hoang and Kanemoto 2021).[38] On the one hand, the dome display might be seen as the government's contribution to the preservation of the cloud forest. On the other hand, it uses damage from climate change to perform a kind of mastery.

Sociologist Catherine Wong points out that although ecological modernization theory has generally tended to promote a move away from government control, its popularization in Singapore and other Asian countries indicates an alignment with a "greater centralization of planning and consolidation of power in state apparatuses and policy elites" (2012, 100). Singapore, with its well-known one-party government, champions this type of eco-authoritarianism.[39] As climate change and many environmental issues become increasingly uncontrollable, the kind of conservation/solution/futures that are subtly recommended by Gardens by the Bay and its domes are those of increased control where—with the right leadership and financial and technological resources—we can safeguard the things that really matter, or reliably produce a particular type of nature, and an aspiring future. Paradoxically, in the midst of the growing uncertainties of climate change, to demonstrate an ability to control the climate and to enhance the existing clean, green image, and an idealized place to live and invest, becomes increasingly more important and valuable. If Biosphere 2 was a life support system for humans, the bio-domes of Gardens by the Bay are a life-support system for authoritarian nature, and also a life-enhancing system for capitalism and eco-modernism.

Singapore's status as the model for sustainable urban living has been cemented by the imposing Supertrees, the bio-domes, and the skyrise greenery that transforms the city with vertical gardens. These feature as technologically sustainable solutions to the tensions resulting from intense urbanization, economic development, space limitations, and environmental degradation. As a result, the urgency of climate change that the Cloud Forest and other similar projects seek to convey is not only buried in a spectacle of authoritarian nature, but their existence and "effectiveness" seem to further validate the need for human control in a world that is threatening and uncertain. To enter the urban future of *Planet Earth II* seems to evoke a reductive yet alluring eco-modernist future, in which technological solutions promise a clean, green urban future that is ultimately an imagining of human control of all elements, secure and financially profitable.

As I wander through this futuristic urban space, in which many green spaces are made possible through social engineering and land reclamation that rely on imported sand, and as I admire the botany in conservatories without the nuisance of mosquitoes, it becomes difficult to dwell for long on the message of biodiversity loss and environmental threats that the architects wish people to absorb while visiting the gardens. Importantly, the irony of self-enclosure implied by the dome is only made possible through tourism, cheap foreign labor, and incessant control over nature and humans. Nevertheless, even this can only continue for so long. When I was in Singapore, my friends complained about the aberrant heavy rainfall: "Weather in Singapore used to be straightforward. There is a clear pattern of wet and dry seasons. July is supposed to be the dry season. But I suppose there is a silver lining. The rain and wind dispersed the haze that choked us in the past." As mentioned in the introduction, the annual haze is a result of the deliberate burning of mainly oil palm plantations and a product of the thirst for biofuel in Indonesia. It is also, in part, guided by the belief that biofuel is more environmentally friendly. For example, researchers have argued that the narrow focus on supposed energy security and carbon emissions has led to policy reforms that promote alternatives like palm oil as an environmentally friendly biofuel despite its negative environmental impacts on the rainforest reserves and wildlife habitat in Indonesia and Malaysia and around the world (Wong 2012; see also Jeswani, Chilvers, and Azapagic 2020). The haze caused noticeable deposits of a wide range of organic and inorganic compounds in Singapore's ecosystems, causing biologists to remind us that "The influence of regional land use changes on Singapore's climate shows that 'no ecosystem is an island'" (Tan and Abdul Hamid 2014, 273). While the pollution certainly has areas of concentration, it has no hard boundaries.

Meanwhile, during my visit to various buildings with award-winning skyrise greenery, the staff members of NParks told me about the unexpected weeds popping up among some vertical greens.[40] They explained their excitement over this unfamiliar discovery in Singapore's well-managed gardens: "There are scheduled maintenances attending to these skyrise gardens. For some reason, some patches were not cleaned up in time. For those ones, there have been reports of more frequent appearance of butterflies. It is not clear yet if they are related, but we are thinking to conduct a systematic study to understand possible correlations."

The haze that knows no borders and the resistant weeds suggest the limitations of control. Underneath the domes of Gardens by the Bay, it may not be immediately obvious that something is causing havoc, yet authoritarian nature ultimately relies on high maintenance to continuously resist the rhythms of many nonhumans. Consider the scale of work required for the garden to function, the constant need to change the plant displays of the Flower Dome to sustain its appeal to visitors, the energy required to keep the Cloud Forest cool, and the reliance on international tourism that itself contributes to significant carbon emissions. The question becomes, who is being controlled and at what expense to support a perceived sustainable model that requires persistent maintenance?

With exacerbated effects of climate change, many more cities are aspiring to become a city in a garden or something similar. In contrast to many cities and countries that are playing catch-up by introducing a green hue into their urban environments, the trope of garden has been fundamental to Singapore's nation building and development since its independence. It has been actively and "creatively" evoked, incorporated, and negotiated. At the same time, although the garden has been used as a way to order meaning in human society, it has also long been backgrounded and instrumentalized. As the city actively cultivates a green and orderly present and pursues a "greener" and more controlled sustainable future, gardens, plants, and some people are simultaneously foregrounded and disavowed.

From the planting of that first Mempat tree in 1963 to the building/ growing of the Supertrees, from the island-wide roadside tree planting to the aggressive promotion of skyrise greenery, the state's emphasis on sustainable development continuously aims to "overcome resource constraints," make economy greener and grow green economy (MEWR and MND 2014, 46). As urban sustainability is increasingly tethered to capital and wealth, perhaps I may borrow Catriona Sandilands's sharp remark: "We are being gardened, not just in a metaphorical sense, but also quite literally" (2019, 6).

As this chapter has shown, authoritarian nature expressed through gardens places a schematic control on humans. It not only produces a kind of modern citizen but helps to produce and prioritize a narrowed frame of sustainability. As urban planning is increasingly oriented toward a sustainable and greener future, attending to the analytical lens of authoritarian nature clarifies these complexities and conflicting perspectives. Furthermore, as it highlights how forms of oppressions are entangled (and enabled) and makes visible the deep-rooted and sometimes normalized dominance on human and other-than-human inhabitants, it also points to ways to challenge and subvert some taken-for-granted greening practices.

How might we enact a different urban imagination from this controlled and controlling garden narrative? The alternative future that Singapore and the world needs to reimagine more seriously is neither a romanticized or human-centric green urbanism nor a singular enhanced capital- and techno-centric one. Rather, we need to open up what authoritarian nature keeps closing down. The new urban narrative cannot be a mere enactment of the human/nature dualism, where the only modification is one of recruiting plants into an eco-modernization narrative. As this chapter has demonstrated, the garden construct and the particular way it has been reimagined have enabled a dominance over human and nonhuman alike, casting serious doubt on whether the "garden" is a helpful figuration for the kind of relationships that are needed at an urban or national scale.

At the same time, control of the plants and humans is only part of the story. The fugitive seeds that arrived with the migrant workers—who have cultivated the land to come and have likely already been forced to leave the city-island—would continue to grow at the reclamation sites across the island, resisting a singular way of naturing and being "natured." What are the ruderal relations that they may forge or have already forged with other human and other-than-human communities? The garden may also offer other possibilities. For instance, Plumwood proposes interspecies gardens that are "designed to provide mutual agency and mutual benefits for a mixed community of humans and non-humans" (2005, 7). In this context, both gardens and people are in the making, are co-making each other, materially and discursively (Hustak and Myers 2012; Myers 2017).[41] From this perspective, the garden or gardening may cultivate and shape humans' agency through ongoing work and care (see Ginn 2014). This kind of reimagining work requires attending to the uncontrollable human and nonhuman and recognizing other forms of nature in its urban fabric.

At the heart of this chapter, my emphasis is on how we imagine more-than-human communities—and these imaginings matter profoundly. The key is to decenter and undermine singular narratives and forms of control. Key aspects of later chapters examine how authoritarian nature and its dominant imaginary can be resisted by certain nonhumans and humans. Indeed, this is the dominant framing within which all efforts to reimagine and produce the more-than-human city must take place.

INTERLUDE: WHO HOLDS MY NAME?

When I was young I learned
the names of our ancestors were stored
in the trees.

I
Farmed in a seedbank, urban trees
do not meet other trees

before moving to the *garden.*
They become the lungs of others,

their bodies weary
from the distance they've travelled.

Their roots are cold, suspended
in these concrete grids.

And when we decide, we fall them
with chainsaws

to make space, to make roads
to make homes, to make up other trees
to make a piece of future.

Urban trees do not meet
other trees. They live

an instant life; they hold
no memory.

II

My ear overworked from trying
to catch up with

accents, thick
like mud. Foreign signs

make me stutter.
In this city, plants are my guides;

they teach me to trace
my rhythm; to remember

to gather my memory; to become
with thousands of other lives.

In the night, I drag
my *half* body out
to meet the trees.

We embrace
to come to ground.

We breathe through
one another   *inversely*

our limbs,
roots knotted.[1]

how

have

they

~~erased~~

our

past

Who

holds

our

2

THE INVISIBLE TIMES

This was my second visit to Lorong Buangkok, the last Kampong on the Singapore mainland. Since the 1960s, Kampongs (meaning village in Malay) like this one have been associated with an old, unhygienic way of living. They have been progressively cleaned up, replaced with public housing and other kinds of urban developments. Most residents were moved to public housing flats built by the Housing and Development Board (HDB). On my first visit, I was challenged by being unable to distinguish the line between the public path and people's homes. This time I felt more confident walking through the unruly vegetation and houses dotted among it. I was here to see Sophia and Sumit,[1] some of the few Singaporeans who still live in the Kampong. All their children and Sophia herself were born in the same house they are now living in. It was built entirely by hand by her father. They had lived for a brief period in an HDB flat until her mother became ill. The family came back to look after the ailing parent and have remained.

We talked about how the village had been much bigger in the past, what constitutes a sense of belonging, and their reluctance to move into a housing flat. The long-term Kampong residents also did not try to disguise the fact that immense efforts are required to maintain their houses:

The zinc roof needs to be replaced every two years; else it would start to leak. It's getting harder to find the material in Singapore. The place also needs to be repainted each year, inside and outside to prepare for the new year. Some utilities we pay each month is actually higher than living in an apartment.

I was particularly interested in how their children coped as all their friends at school lived in flats. Sophia said: "We are the only family that has internet in this block. For the kids to stay, you got to give them something in return."

She laughed, then added: "They are good kids. I think they are a bit more patient than their cousins and friends, perhaps this house and our way of life made them so."

When I pointed out the abundant potted plants placed around their home, Sumit said they belong to his brother-in-law, who now lives in a public housing flat but often comes back to help maintain the place and look after the plants.

He seems to take more interest in plants after he moved to an HDB flat. We do not want to move. My wife can't adjust to apartment living . . . doesn't like those walls. There is no Kampong spirit there . . . like, how our neighbours just came to help when a big tree fell last week. . . . But sometimes I am a bit worried about the safety of my family when I am away. . . . No no no, it's not people. The snakes. I am concerned about them getting into the house at night.

As Sophia looked at a development site of multiple high-rise buildings right next to their home, her gaze moved to somewhere far off and invisible to me.

We had a huge fish pond, and prawns . . . just next door. Now, all constructions. . . . They bring so much dust and noises . . . I don't think we will be able to live here for long.

While Sophia and Sumit explained to me their unwillingness to be confined to apartment living surrounded by walls, I walked around and discovered new aspects to the place. I began to realize how words sometimes fail us and how attached we can be to the environment around us and how difficult it is to express our attendant emotions. The site of a huge fallen tree had now been cleared and a hand-written sign "Danger, Falling Branches" removed. There was a large cluster of bamboo that threatened to collapse on the side of the house. Sumit said that as soon as I left, he would need to find a way to make them more upright. Here, the lines between a physical "home" and its surroundings are blurred. It is hard to discern where the

2.1 This last pocket of Kampong in Singapore is surrounded by new housing development. *Source*: photo by author.

boundary of their home might end and that of their neighbors start. At
their home, it is perhaps the diverse needs of living and engaging with the
environment that help them to feel rooted, creating and sustaining a sense
of belonging.

To the rest of Singapore, Lorong Buangkok (see figure 2.1) has become a
living museum, a way of life that is trapped in the past and a place where
time seems to be out of joint. Yet as I greeted Sophia's mother, an eighty-
year-old lady who has called this Kampong home for almost her entire
life, chatted with their teenage daughter, and took photos of the aging
house that demands tremendous care and patience, it didn't seem as if
their lives were any less lively than those of any others. But walking around
in Lorong Buangkok, I know my steps follow a different rhythm and my
heart is lighter.

During the rest of my time in Singapore, I spoke with a range of other
local people about the "last" Kampong. For many of them, the Kampong
takes up too much space, and the maintenance of the Kampong home is
exceedingly time-consuming and inconceivable in a modern city. In sum, it
is a way of living that belongs to the past. In discussing urban development,

attention is often drawn to the increasing pressure and demand for space. For Singapore, a country that can no longer expand but only fold in, and grow upward or downward, issues and tensions arising from the scarcity of space are magnified. As I look at the photos of Sumit and Sophia standing in front of their home against the high-rise buildings under construction, with a large surrounding area they need to attend to and share with other human and nonhuman inhabitants of the Kampong, it becomes clear that their resistance to moving into a HDB flat is not only about spatial confinement. It is also an unwillingness to conform to a single way of inhabiting time directed by the never-ceasing development regime: to streamline and to be productive, to make time count.

Colonization of a place is achieved as much through dominance of a particular way of time-making and futuring as it is about spatial relations. As Barbara Adam notes, "Nature, the environment and sustainability, however, are not merely matters of space but fundamentally temporal realms, processes and concepts," hence "[w]ithout a deep knowledge of this temporal complexity . . . environmental action and policy is bound to run aground, unable to lift itself from the spatial dead-end of its own making" (1998, 8). More specifically, urban humanities scholars Dana Cuff and colleagues point out that it is crucial to "consider the need to rethink *who* can possess the future, and *how* such speculation ought to happen" (2020, 175, italics in original).[2]

This chapter explores how the making and representation of time shapes and is shaped by urban development and its residents, their layered entanglement with development and heritage, with remembering and forgetting. Here, many things, humans and not, are my guides to open up a new inquiry into the often-normalized displacement—uprooting and relocating of multispecies urban residents—and its generational social, cultural, and environmental impacts. Each thread evokes different sentiments with a complex desire for both new and old, an accelerated life and static mode of living, marching forward and remaining constant. Responding to these complex and messy challenges, I propose the concept of *double erasure* in an effort to encapsulate the ways in which the removal and/or alteration of physical sites may bring about further erasures of memories and culture. I ask how an insidious and interrelated process of erasures may disrupt the relations between the remembrance and the environment. At the center of this chapter is an insistence on the need to pay attention to the effect of an enforced unified time on humans and other-than-humans, as well as the possibilities opened up by learning to attend to other ways of life and temporal rhythms.

In the first two sections, I look through the lens of my visit to the HDB hub at how the island-wide public housing development program and the demolition of Kampongs is legitimized by a future-oriented temporality, and the ways in which the concept of heritage is used to construct a certain natureculture past that enhances the national narrative and the present. I am interested in how the reduction of temporal diversity impacts and disrupts other forms of diversity, both cultural and biological. In the third section, I focus on how these bio-necro-material disappearances give rise to double erasure, creating a kind of amnesia that inhibits us from imagining other possible worlds. Through exploring various ways of remembering and participating in a local program run by the Ground-Up Initiative, a Singaporean social enterprise, the final section reimagines how an active mode of remembering may subvert a forced synchronization underpinned by a singular, linear way of inhabiting time.

MARCHING INTO THE FUTURE

Although it appears that objective clock time is now ubiquitous, its forceful and at times violent process of adoption continued until as recently as the early twentieth century. In the great satirical novel *The Time Regulation Institute* (1962/2014), Turkish writer Ahmet Hamdi Tanpinar depicts a frantic period in which universalized mechanical time penetrates all aspects of society. This is depicted as part of the radical reformation of the Ottoman empire into a modern state. Mustafa Kemal Atatürk, the founding father of modern Turkey, believed that in order to catch up with the Western way of life, the country must first align itself with the modern way of recording time, clock time. Hence, in 1926, the Gregorian Calendar Act replaced the Islamic Calendar, and Western time was enforced throughout Turkey. In the novel, Tanpinar imagines an institution that capitalizes on this transformation by policing time. The mechanism involves fines for any clocks or watches that are either running fast or slow and therefore not synchronized with a modern clock. In real life, any unruly ways of inhabiting time that do not follow the now institutionalized temporal rhythm of society are penalized by being replaced or erased. Of course, the state is at the center of this temporal control. All over the world, as political leaders realized the "heavy economic and political price to be paid for any deviance from the industrial temporal norm" (Adam 2002, 17), they erected clock towers in city centers as an emblem of the new authoritarian eye, through which different modes of life could be observed as they were assimilated by advancing modernization.[3]

In Singapore, Lee started to pursue his dream of turning Singapore into a highly developed state and instituted full industrialization immediately when he gained power in the 1960s. Alongside the vision for a Garden City (see chapter 1), under his leadership, the linear time of modernization tightened its grip through the management of people's homes. This was made possible by one of the most successful state-run public housing programs in the world. Villages throughout the island were demolished and residents of various ethnic groups resettled into unified public housing in the process. Akin to the clock tower overseeing the exercise of an all-encompassing modern time, the construction of public housing, which over 80 percent of Singaporeans now call home, has become the center and symbol of the new island-nation.

In order to grasp the mastermind behind Singapore's public housing program and understand how the modern way of inhabiting time has shaped people's daily lives, I visited the HDB hub where the future is imagined and sold all in one convenient location. The HDB gallery is a multisensory exhibition detailing the trajectory of housing development over the past fifty years, from the Kampong to the futuristic eco-estate, or in the words of the official pamphlet: "from Slum to Vibrant Towns." According to the official narrative, only 8 percent of the population lived in public housing in 1957, while "most people lived in slums and Kampongs which were cramped, dirty and unsafe." The young government faced a great and urgent housing need given the growing population. In 1961, a disastrous fire in Bukit Ho Swee left sixteen thousand Kampong dwellers homeless. Fortunately, within a week, a large proportion of victims were rehoused in new housing flats, signifying a new start to life. By the end of the 1970s, 50 percent of Singapore's population lived in HDB flats. To date, more than one million flats have been built by the HDB. After a brief look into the past, the exhibition quickly moved on to display the master plan of different housing towns. The remaining exhibit centered on the state's efforts and vision to develop sustainable smart living.

After seeing a series of photos and displays illustrating the poor living conditions of the Kampong at the gallery, I turned to various researchers' work for a fuller sense of the past. How did multiracial communities live in the Kampong? How did the residents cope with the transformation of their lives? Details of uprooting and resettling as well as the social disruption over forced synchronization started to emerge. Historian C. M. Turnbull highlights the deep distress caused by the clearance as while it was welcomed by some Kampong dwellers, others who had lived in flourishing village communities "in well-kept houses with gardens, chickens and fruit trees . . . were swept into the anonymity of high-rise flats not to give them a better standard of living but because their land was required for other purposes" (2009, 349). In contrast to the state's historical narrative, historian Loh Kah Seng asserts that public housing development does not "simply aim to meet the needs of growing populations"; rather, social engineering is essential for the ruling elite to create its imagined modern nation-state (2009, 155).

As modern Turkey and some other countries legitimized their political power by enforcing Western clock time and disregarding all other cultural and personal times, Singapore's ruling party, the PAP, cemented its rule by imposing a unified future made manifest in an island-wide housing program. But these two forms of management are not unrelated. In Singapore, the management of housing became a central strategy by which to manage time and vice versa. Sociologist Carol Greenhouse's approach linking temporality to legitimacy provides insight into the workings of this type of temporal politics. In Greenhouse's account, "Social time is not about time passing but about the vulnerability of political institutions to legitimacy crises of various kinds" and is "always plural and always contested" (1996, 15 and 7). In other words, as Anne Brydon succinctly explains: "Representations of time are manipulated in moments of crisis to legitimize institutions of law, politics, and scholarship" (1999, 993).

The early 1960s was a turbulent time in Singapore. The newly formed PAP faced multiple crises when it came to power. These ranged from the resurgence of a left-wing movement and the 1964 racial riots to Singapore's separation from Malaysia. It was imperative for PAP to cement its power quickly. And what would be more effective and quantifiable than "housing a nation" to unite its people? Legitimized by this vision, the Land Acquisition Act, which allowed the state to purchase private land for public use, was passed in 1967, giving the HDB absolute power in pursuing the state's right to pursue urban clearance and renewal. To convince people to abandon their existing homes and move into housing flats, the Kampong and its associated mode of living was powerfully presented as backward, unhygienic, unsafe, and a lifestyle that was "old fashioned and out of place in modern Singapore" (Turnbull 2009, 349). In this way, the Kampong time was deemed to be unproductive and past time, at least in the context of the linear time of modern industrial production.[4]

Returning from the gallery to the atrium of the HDB Hub, I found myself among a group of couples and families as they were examining the future homes available for sale. At the center of the atrium, we were welcomed by a large temporary exhibition featuring the most recent and prominent housing development: Tengah. Dubbed the "Forest Town," Tengah, a new estate under development, is said to represent the pinnacle of HDB's vision of sustainable, smart, and future living. I thought of Greenhouse's proposition that the singular, unified way of inhabiting the temporal realm is essentially a tool to bridge the "incommensurabilities of peoples' imagination, sense, means, and ends" (1996, 211). Indeed, the HDB project was also deployed to manage diversity. In the 1960s, with a diverse population of 75 percent Chinese, 17 percent Malays, and 7 percent Indians and others, the Kampong with its more diverse social rhythms was depicted as encouraging "communal ghettos that impeded multiracial nation building" (Turnbull 2009, 349). Thus, the scheme of HDB flats that imposed a unified temporal pace on a diverse group of people was understood as aiding the integration of citizens "into the social and economic life of the emergent nation-state" (Loh 2009, 154).

Along with resettling residents into apartments and into a modernist time frame, Singapore shifted its focus from farming to labor-intensive manufacturing, and later to a high-technology and knowledge-based economy. In addition to authoritarian nature's efforts to cultivate modern citizens, researchers point out that the layered and entangled national home-ownership program has contributed to the transformation and disciplining of the population into an industrial labor force who are inclined to support the status quo (Chua 2014, 521).[5] Facing a multitude of challenges to its ruling status, PAP legitimized its power through the construction of new dwellings. In so doing, it not only aligned its existence with a singular mode of life that it promoted but in the process reduced or prevented synchronization with alternative ways of living and imagining. Hailed through the ruling elite's own mythology, the great migration from the Kampong to housing flats is officially commemorated as the pivot of the Singapore Success Story, compressing the historical narrative of the city-state to a mere fifty years of modernization.

However, "a life story in linear form can never be personal" (Greenhouse 1996, 209). The consequence of imposing a collective modernist narrative is never singular as temporal diversities are needed to sustain diversities of other kinds. Having been rehoused in HDB flats, there was a marked reduction in the autonomy of families who lost their "freedom to move homes and sublet, rent, build or rebuild their houses on their own terms" (Loh 2009, 154). Moreover, resettling the diverse population from Kampongs to housing flats left significantly different marks on each of the affected communities. For example, while the resettled Chinese in high-rise HDB flats continued to be the majority population, the Malay who used to live in their own Kampongs were dispersed and became "permanent minorities in all housing estates" (Chua 2005, 7). In 1989, the government instituted a strict quota system on each block of public housing flats that "roughly corresponds with the proportion of the three race-groups in the total population" (7). While the quota system appears to be "'fair' and 'rational,'" Chua points out that the consequences are "grossly unequally distributed" between various ethnic groups:

the Malays and Indians are now completely minoritized. This has very significant negative effects on the daily life of these two race-groups. For example, a Malay household will now have greater difficulty in finding an immediate Malay neighbour who would understand all the religion based cultural practices to come to its assistance in case of emergency. The Chinese of course would not face similar difficulties. (2005, 8)[6]

In Singapore, the mass uprooting and relocation from the Kampong to public housing has disregarded the complex spatial (see Hee and Ooi 2003) and temporal relations of these villages.[7] The housing project has become the most powerful and measurable way in which the state regulates time, and manages the pace of life, labor, and the imagined future.

Further, even after the majority of the residents on the Singapore mainland moved to the HDB, various displacement and erasure work continued in the outer islands of the city-state. Urban geographers Creighton Connolly and Hamzah Muzaini (2022) have traced the history of some of these offshore islands and their radical transformation as part of Singapore's envisioning of a modernized future. In their account, since the 1970s, the residents of these islands, mostly Malay and Orang Asli were forced to leave their home of centuries as the islands were turned into sites as petrochemical processing centers, a landfill, or tourism locations. This metabolic and spatial transformation has provided the much-needed capital for Singapore's development, enabling the urbanization and greening work of the mainland (Connolly and Muzaini). Thinking with Jonathan Silver, Connolly and Muzaini suggest that the displacement of the island communities produces a form of socio-ecological violence that is highly unevenly distributed, "affecting particularly Singapore's Malay and indigenous *orang laut* [meaning sea people in Malay] communities" (2). After forced removal, most of these residents moved to Singapore mainland, and into HDB housings, effectively ending their ways of life.[8] Meanwhile, the process of land reclamation, including merging various islands, caused devastating loss of mangroves and coral reef, a point that I will return to in chapter 4.

As the island moves along a determined path of modernization, the inheritance from the diverse past and the intra-related ways of living are being replaced by a seemingly harmonious modern way of inhabiting the present, aligning the stories of individuals with a linear national narrative of progress. In Michelle Bastian's account, linear social time is "not simply *the* time," but rather "a method of attempting to co-opt and/ or exclude diverse others" (2013, 23, italics in original). As varied forms of life have to make way for a singular way of moving forward, their possible futures are "thus dealt with and eliminated in the present" (Adam 1998, 57). Importantly, the practice of uprooting and relocating to accelerate the realization of a desired urban future certainly is not limited to human residents. It is, as discussed in chapter 1, entwined with complex greening and de-greening practices as part of the Garden City Movement. The reduction of temporal diversity through mass displacement ultimately reduces other forms of diversity. As it forgets that one of the key elements of leading to a sustainable environment is to sustain diversity, it in turn inhibits our ability to imagine other possibilities of living.

CURATING THE PAST

In addition to using a future-oriented narrative to manage diversity, temporal control is also managed by evoking and curating a representative and stable past. In 2001, the Heritage Tree/Road scheme was launched. Its aim is to conserve Singapore's mature trees and some of its most scenic and significant tree-lined roads. According to the NParks, the preservation "adds an element of permanence to the landscape and ultimately contributes to Singapore's sense of identity, history and continuity."[9] There is a stringent qualification criteria for heritage trees. A tree must have a girth of more than five meters and carry botanical, social, historical, cultural and/or aesthetical value. Or, as a local succinctly told me, a mere old urban tree without a story can hardly be eligible for heritage status. As of 2016, approximately 260 trees and five roads were registered as heritage tree/roads, and hence are legally protected.

As I wandered around in the city looking for these trees that supposedly celebrate Singapore's natural heritage and represent the relationship between trees and humans, I noticed how isolated they were. Standing alone or among other newly planted greenery, the monumentalized trees along with the memories they are supposed to preserve appeared strangely distant. The highlight of my search was an encounter with "the Avenue of Heritage Trees" on Connaught Drive. While other heritage trees needed more introduction to draw my attention, the twenty-two majestic rain trees suggest a strong sense of continuity. Walking underneath their gracious canopies, a smile lit up my face.

As one of the most urbanized places in the world whose success is contingent on a constant remaking of its image, the anxiety of rootlessness suffered by its human inhabitants seems to be partially soothed by a sense of permanence suggested by the deep roots of these majestic beings. However, the crafting of the past is crystallized in highly selective conservation practices toward various types of greeneries. Although Singapore has few legally protected green areas (mainly primary forests) despite being a City in a Garden, it aims to be a leader in the development of urban biodiversity while continuing its "business as usual" agenda.[10] As a result, it is essential for the state to present and preserve a "pristine" past, or in the government's words, "a significant and representative segment of our native ecosystem," through meticulous planning (Urban Redevelopment Authority 2014). Secondary forests with supposedly less biodiversity and comprised largely of exotic species are therefore deemed unworthy of preserving (Yee et al. 2016). In contrast, the very few protected reserves serve a dual role. They provide a genetic seedling resource to maintain the native flora and fauna and a representation of authentic nature from an imagined past.

In Matthew Gandy's account, the urban ecological imaginary still carries a "dichotomous" and "neo-romantic" undertone that excludes nature from many deep-rooted social cultural issues (2006, 62). Heeded by this problematic thinking of nature, Gandy argues that the transformation of the urban seems to tend toward an imagined "equilibrium state" guided by some purported nature parameters (61). Indeed, using nature as a metaphor for a state of balance or to evoke an imagined past has been central to its impacts on various aspects of urban development. These discussions are particularly illuminating when engaging with the issue of heritage and urban biodiversity conservation in imagining a sustainable city.

Situated in constantly shifting surroundings, these heritage trees and "pristine" primary forests form a contrasting backdrop to the fast, transient urban life. Intriguingly, Kampong Lorong Buangkok, standing among rising housing development projects, is perceived by some as serving a similar role. While I was sitting on the porch of Sumit and Sophia's house, quite a few people wandered around taking photos. Among them was a young couple dressed in decorative traditional clothes followed by a photographer. Sophia told me that outside visitors are no longer an uncommon sight in Lorong Buangkok. The couple were perhaps having their pre-wedding photos taken in the "last" Kampong to "borrow" a sense of permanence in a futuristic city-state. Here, although the Kampong, nature and "modern" people all momentarily share the same physical space, there is an obvious inconsistency in the temporal realms they reside in. Rather than being accepted as an alternative way of life, the Kampong and its rhythm of life are generally perceived as of the past or the other, in contrast to the dominant reality.

Anthropologist Johannes Fabian proposed the term "denial of coevalness" to describe a persistent and systematic tendency to reject the contemporaneity of a deemed other, suggesting a refusal to share a common time (1983, 31). In urban and nation building, these "rhetorical elements of temporal distancing": "here and now" and "there and then" (Fabian, xi and 103) are fully deployed. We see this with the Kampong but also with the heritage tree serving as a metonym for an imagined permanent nature, a soothing agent for modern anxiety or a validation of certain human histories. With Lorong Buangkok's rising popularity as the last Kampong on the Singapore mainland and the increasing number of tourists, its residents' way of life, deemed backward and unproductive as opposed to that of HDB flats, is not only under threat by modern development but also at risk of being preserved as a static past, a photo opportunity of a heritage site among modern buildings. Either way, despised or romanticized, there is little intention to seek any genuine form of coevalness with this mode of life in the present, thereby reducing the possibility of moving together into a shared future.

In alignment with the heritage trees and primary forests that are said to help Singaporeans feel rooted in their homeland, the government has increasingly focused on cultural conservation. They aim to create a sense of attachment to familiar places and to maintain "identities and social memories" in hope to stem outward migration (Henderson 2011, 50) and an eroding sense of national identity. Some of my time in Singapore coincided with the annual heritage festival organized by the National Heritage Board. A range of programs were promoted through local news channels, newspapers, and social media. Carefully prepared catalogues illustrating the history of various heritage districts were distributed throughout the island. Historian and geographer David Lowenthal writes, "History explores and explains pasts grown ever more opaque over time; heritage clarifies pasts so as to infuse them with present purposes" (1998, xv). Indeed, what to preserve or not is constantly being negotiated between social, economic, cultural, political, and environmental needs (van Dooren 2017). My experience so far has taught me that far from being based exclusively on length of time, heritage is caught in a temporal knot in which the past, the present, and the future coalesce.

Around the globe, heritage planning is often directly linked to tourism. In the 1980s, concerned that Singapore was losing its "oriental mystique and charm" (Ministry of Trade and Industry 1984, 15) and might be overlooked as a tourist destination due to two decades of "demolish-and-rebuild" (Yeoh and Kong 2012, 128), the government started a heritage program focusing on ethnically prominent areas such as Chinatown to contribute to its overall repositioning as "a unique mix of heritage and modernity" (Centre for Liveable Cities 2015, 63). This preservation was later criticized as "a version of multiculturalism based on consumerism . . . as well as symbolic representation" (Yeoh and Kong, 146). From a political perspective, the conservation of cultural heritage is harnessed for nation-building (Yuen 2005; Henderson 2011; Low 2017; see also Kong and Yeoh 1994). With a complicated colonial past and a compressed modern history, the government demands a strong adherence to multiracialism as part of its national identity in order to "cement cohesion across racial divides and consolidate its own position" (Henderson, 51). Guided by this, almost half of Singapore's national monuments "are allied to religion and are of matching architecture" to ensure the visible presence of multiracial roots (Henderson, 52). Heritage and tourism scholar Joan Henderson further notes that while most of these preserved religious buildings still perform their original functions, other preserved buildings have often been converted to commercial use if they did not have an original religious function. The original facades of these buildings were generally restored for their visibility in an effort to deliver something akin to an authentic historical image.

As part of this process, the Singapore River area was preserved through a complete reinvention with conserved architectural facades in keeping with the state's objective of transforming the river into a world-class site for entertainment, recreation, tourism, and culture. During the process, the traditional wet market and public housing blocks were torn down. The once spontaneous or more accessible cultural and artistic activities such as traditional Chinese Opera or the buskers' festival have been replaced by organized events (Chang and Huang 2005). These have been criticized by some as "highly contrived and commercialised" (Chang and Huang, 275). There is little remaining that alludes to the river's rich and diverse past and its continuation into the present (Chang and Huang; Rose 2021). The severed connection between residents and the river would later have a profound impact on people's imagining of water. People's relationship with rivers and water, or put more accurately, their disassociation would become a key barrier to the country's call for water conservation. It has also cultivated the conditions that have given rise to a reliance on technological water production.[11]

In Singapore's future-making project, multiple pasts, humans, and nonhumans have been dislocated in favor of the careful crafting and preservation of a particular version of the past that enables a certain future. In selecting heritage sites, their value is conditional on their length of existence and association with human history, as well as whether their past can be distilled in a way that assists the present. The liveliness, agency, and deep past of heritage is reduced to representational forces, serving as a backdrop or providing a sense of history to a transient modernization narrative. Having carefully admired the conserved cultural and natural heritage sites, it is time to recuperate and continue the modern way of life dictated by the linear path of progress.[12] More insidiously, these conserved buildings with their intact facades and heritage trees work to assure us that everything important from the past is being retained.

THE DOUBLE ERASURE

In their reply to "An Eco-modernist Manifesto," an article that celebrates the age of the Anthropocene in which humans are said to be taking control of planetary systems, Rosemary-Claire Collard, Jessica Dempsey, and Juanita Sundberg condemn the amnesia of eco-modernist approaches in the following way: "If there is a singular trait to describe *An Ecomodernist Manifesto*, this is it. Amnesia. In two registers. First, amnesia about the deeply uneven and violent nature of modernization. And second, about the struggles that have underpinned every effort to alleviate inequality and violence" (2015b, 227). Here, Collard and colleagues warn against the eco-modernizers who seem to have forgotten that the many violent processes of modernization have resulted in a world of deep injustice. Echoing this forgetting, other kinds of amnesia are developing in Singapore's accelerated realization of its desired urban image. This is assisted by an active

erasure of the built and ecological environment and its associated past. The
local poet Alfian Sa'at (2008) offers his diagnosis of the modern city-state
in these stunning lines:

The patient, born in 1965,
suffers from a history of
amnesia of unknown onset
and duration.

The prognosis is uncertain.

How does modernization/eco-modernization normalize forgetting? Reflecting on the island's wide displacement of the human and nonhuman discussed earlier, and inspired by Deborah Bird Rose's (2004, 2005) notion
of "double death," I have developed the concept of *double erasure* as an
approach through which to explore the process and relations of removal
of the built/natural environment and a second erasure of associated cultural ecological relations and their memory.[13] The entwined disappearances
culminating in this form of amnesia sterilize the generative ability of the
past (metaphorically and literally) and close down alternative perspectives
for the future. Under the regime of modernism/eco-modernism, prominent
questions such as what might be sustained into the future, by whom and for
whom, are often obscured and narrowed.

It is important to pay attention to the impact of double erasure when
meditating on the narrative of sustainability. This is especially so as the
notion of sustainability and the act to sustain is positioned at the nexus
of the entanglement of the past, the present, and the future. To illustrate
these interdependent erasures, an examination of the clearance of the
Kampong is useful. The removal and demolition of the villages also gave
rise to another prominent disappearance: the ethos of the *Kampong spirit*,
broadly explained as one of helping each other out and extended kinship.
Although village life has been looked down on during the process of modernization, the Kampong spirit that emerged and evolved from the quintes-
sential Kampong way of living is increasingly considered highly desirable
by the government and so needs to be revived. So much so that one of
the new developments is named Kampung Admiralty and is described as
"a modern interpretation of 'Kampung' living, as one can dine, socialise,

shop, and receive healthcare services, all under 1 roof" (HDB 2023). Even the greenery in the new estate was specifically selected for its links to Kampong life (Heng and Chua 2014).

This attempt to manufacture the Kampong spirit from certain identified ingredients is a gross simplification of the relationship between memory and environment. Following the removal of pine forests in Australia, Katherine Wright describes the displacement of her "environmentally embodied" memories, a term she borrows from Mark Rowlands. The intimate relations that Wright suggests between memory and place are not straightforward or linear. Rather, she refers to memory as an ecologically involved process that connects "along lines of shared earth and experience" (2012, 124). Indeed, the assumption that the Kampong spirit can be easily re-created fails to recognize that this ethos, and community relations of helping others, is grounded in a profound more-than-human entanglement between Kampong dwellers and their surroundings, and is invested and performed in a range of material practices. In Lurong Buangkok, there are no clear boundaries between Sophia and Sumit's home, the neighbors' home, and the wider surroundings. The emergence of the Kampong spirit is a layered process in which residents experience and "intra-act" (Barad 2007) with the others, underpinned by a deep sense of care and appreciation of cohabitation. Thus, the demolishing of the site and the environment ultimately erases this nourishing relation, exemplifying the current and latent implications of erasure work.

Double erasure continues to have an impact in the removal of the built environment. Despite selected preservation programs, historian Lai Chee Kien (2018) laments that heritage buildings in Singapore have been and continue to be impacted by new developments, including transport infrastructure. In 2004, the old National Library Building that carried generations of Singaporean memories was torn down to make way for the construction of the Fort Canning Tunnel to ease traffic congestion. The announcement of the planned demolition caused furious public objections and outcries. Alfian mourns the library and other lost buildings in his poem:

So these bricks will be torn down
And books will still not have learnt
To spread their feathers and fly
Like pigeons from a shaken tree.

So this balustrade will be dismantled
Perhaps reassembled somewhere else—
A conch paperweight by my bed is a beach.
Each hour from a postcard Big Ben chimes. (2008, 45)

The imagery of the demolished library as a falling tree resonates deeply
with the concurrent deforestation on the island. In the poem, Alfian
describes the erasure of the library and the subsequent vain attempt to
recycle the memory of the library in other forms, for example, through
reassembling the dismantled balustrade. Cultural theorist Joanne Leow
(2011) gives her heartfelt reading on the poem: "the reduction of this
collective knowledge and shared space to mere souvenirs and postcards
is shown to be futile. . . . Alfian notes this as an absurdity and affront
using the surreal images of 'a conch paperweight' as 'a beach' and Big Ben
chiming from a postcard." In the final stanza, Alfian writes:

On my shoulder. I opened that book,
Expecting a cry for help, a refugee's plea.
What I found instead was this poem
That did not know how to end. Only when. (46)

Here what is particularly alarming is that there is no "cry for help" in Alfian's
recounting of the erasure of the library. There is only an inconclusive ending haunted by a lingering sense of loss and "the eventual acceptance of the
library's demise creates a certain paralysis in the collective memory" (Leow
2011). In this case, the kind of acceptance, or more precisely an inability
to remember, manifests the work of double erasure; that is, the result of the
ongoing erasure of the environment leaves behind a fragmented existence
for humans and nonhumans and an erased past and liveliness.

Double erasure haunts the dead as much as the living. In 1978, the
then minister for law, environment, and science and technology, E. W.
Barker, announced that all private cemeteries would be "acquired as and
when required for development" (Hansard). Many cemeteries have been
lost since, including the primary Bidadari cemetery, from which 58,000
Christian and 68,000 Muslim graves were exhumed to make way for new
public housing (Chow n.d.). To meet current and future traffic needs,
the Land Transportation Authority of Singapore announced in 2011 that
a new highway would be constructed cutting through the Bukit Brown

Cemetery, the home of more than 100,000 graves (some as early as the late nineteenth century) and a mature secondary forest with rich biodiversity (Huang 2014).

Speaking to *The Guardian*, Darren Koh, an associate professor of law and a volunteer at All Things Bukit Brown, an activist group that endeavors to save the cemetery, said, "The whole cemetery culture is already gone. . . . The idea of worshipping or honouring the Earth deity first before you go to the grave, it's gone." *The Guardian*'s report further notes that the increasing "estrangement from the 'culture' of cemeteries has led to burial being seen as a waste of space in Singapore" and more people opt for "cremation in the first place" (Han 2015).[14]

During a conversation with Darren, he told me: "We know we can't save everything. . . . For us, it is to let more people know this cemetery is a living heritage . . . from various burial customs, the history of the materials used in building the tomb, to the inter-generational connection and rich biodiversity of the forest. . . . After knowing all these, people can decide themselves if the place is worth conserving." Darren went on to say, "If my primary school no longer exists, my playground no longer exists. The home I grew up in is gone, the road I lived on is gone. What is there left for me in Singapore?" Walking in Bukit Brown Cemetery among the living and the dead (see figure 2.2), what Darren said to me kept coming back: Singaporeans shouldn't be forced to choose between their grandfather's grave and their grandchildren's home.

The cemetery, of course, is never only the home for the dead. As well as being a crucial cultural heritage, plant ecologist Ingo Kowarik (2020) reminds us that urban cemeteries are often biodiversity hotspots because of the rock formations of gravestone and undergrowth. The rotten woods in different stages of decay also offer key habitats for a wide range of other-than-human residents, sustaining death into life (Kowarik; Rose 2004). As in the erasure of the Kampong and its environment, and the subsequent Kampong spirit, years of exhumation and relocation of the ancestors' tombs is accompanied by a gradual disappearance of associated memories and the whole burial culture.[15] Eventually, it results in a changed view of the connection between the dead and the living.

In *Moving House* (Tan 2001), a documentary that tells the story of the Chew family in Singapore, one of the 55,000 families forced to exhume

2.2 Bukit Brown Cemetery is under threat to make way for development work. *Source:* photo by author.

the remains of their parents and transplant them to a columbarium, the voiceover reveals the latent disastrous impact of double erasure: "For Singaporeans, moving, rebuilding, resettling, whether voluntary or involuntary, is a way of life. And the dead are not exempted." Akin to the ending of Alfian's poem, the removal of the environment in which cultural relations have long been cultivated results, for those affected, in an inability to remember or to resist the forceful forward momentum. If the future is imbued with possibilities, double erasure becomes a force to interrupt, and to make the past and the future still and infertile. Returning to the stories of the outer islands of Singapore, Connolly and Muzaini point out that not only is there no longer "any trace of the villages," the radical alteration of "spatial forms and cultural landscapes" of the islands has seen these islands now largely absent in most Singaporeans' mentalscape (2022, 15). Here, double erasure continues its work of amnesia on multiple fronts.

Social anthropologist Paul Connerton suggests that there are at least seven types of forgetting. In the context of the Singaporean state, it is particularly helpful to think with what Connerton describes as one of forgetting "that is constitutive in the formation of a new identity" centering on what can be gained from disgarding the memories (2008, 62–63). In Singapore, the primary discourse of sustainable development, or debate on the necessity for conservation, is predicated on the need to sacrifice or actively forget some aspects of an individual's past for the sake of future generations. Here, the state's deliberate patterned forgetting aims to provide "living space for present projects" (Connerton, 63), and moreover, for an imagined future. Yet, this type of forgetting narrows the concept of connection and continuity to a mere linear form and ignores the "interdependent, contingent connectivity" (Adam 1998, 46) between times and generations and a wider more-than-human world. The new identity that arises from this kind of amnesia seems to be a troubled one. This communal "acceptance" of a transient way of living that breaks down relations between the past and the present, and between the present and the future, fundamentally destabilizes the ground of meaningful discussions on sustainability or heritage preservation.

As double erasure eliminates physical presence and the attendant entangled more-than-human relations and memories, amnesia in an ecomodernist sense opens up space for a new script or, in other words, a new identity. To return to the HDB Gallery, the exhibit featuring the Tengah Forest Town, a prominent eco-smart housing estate under construction, does just this type of work. The major attraction of the new development includes the creation of a one hundred-meter-wide forest corridor as part of the estate, projecting futuristic forest-like living (HDB 2016). To carry out its current plan, the housing project would reduce the existing Tengah forest from seven hundred hectares to fifty hectares, accompanied by planting other kinds of greenery that are, according to the local conservationist Ho Hua Chew, "not attractive to forest wildlife" (Juanda 2016; NSS 2018). The goal, of course, is not to remove nature entirely. In fact, the aim is, as I have argued in chapter 1, to actively incorporate nature, but of a particular kind that is spatially manageable among housing and contributes to the creation of an imagined future of sustainable housing.

But the development of the new estate is also more than simply a case of recruiting nature. As the current forest and scrubland are cleared, double erasure not only erases the forest but at the same time eliminates the temporal and ecological relations and existing functions of the secondary forest as a refuge for some animals (some were pushed out of their original habitat). With the layered and interwoven displacement of cultural and natural heritage and subsequent forgetting, the intra-acted erasure ultimately helps to legitimize and normalize the persistent modification of landscape, streetscape, bio-necro-scape or anything that does not fit a certain path of progress.

As the landscape is constantly transformed for development, its complicated past erased or forgotten, the remainder is easily positioned as merely a dispensable patch. Meanwhile, the new mode of living consisting of forty-two thousand new homes with smart sustainable features positioned among selectively preserved and newly planted greenery crafts a new imagining and aspirational way of being "At home with Nature." Imagined and enacted in this way, the new HDB development, framed within the narrative of sustainability, produces a new baseline for eco-living for the generations to come. In so doing, the eco-modernized future that has obliterated the life forces of many, appears to sustain life using a controlled and designed generativity. Perhaps this is how eco-modernist narratives are constructed, as if there were no history. In modern times, we are always ready to forget.

As double erasure is ultimately a layered process of forgetting, I would like to touch on its supposed antidote—remembering. In many ways, the battle with climate change and the resistance to eco-modernism are increasingly related to our ability to re-collect and re-member.

HOW DO YOU REMEMBER?

In her discussion on the interrelationship between memory, nostalgia, and the city of Singapore, Leow (2010) suggests that the younger and older generations of Singaporeans have different longings for the past. Unlike the older generation's attachment to the lost *Kampong*, the younger generation in Singapore is "nostalgic for unachieved authenticities and truths in the country's historical past" and speculates about new connections and possibilities other than the national narrative of progress (Leow, 119). These seemingly different longings are in fact the result of being deprived of generational memories. For the new generations who suffer from erased natural/cultural/built environments as part of double erasure; what has been forgotten and in turn what needs to be remembered are no longer clear.

In recent years, Singapore has entered a "memory boom." In 2011, the Singapore National Library Board launched the state-sponsored "whole-of-nation movement," the Singapore Memory Project (SMP), to "capture and document" memories related to the country (Pin 2014). After analyzing the structure and contents of the project, historian Kevin Blackburn points out many important missing moments of Singapore history such that "the tumultuous and contentious political events of the 1950s and 1960s" are not mentioned in anyone's recollections (2013, 452). Others suggest that SMP primarily focuses on the individual's "memorability of growing up and witnessing change and development together with the progress of the nation-state," from third world to first (Liew and Pang 2015, 558).

Furthermore, the SMP has been criticized by some researchers and members of the public for romanticizing the past, using nostalgia as a political tool to soothe modern anxiety (Blackburn 2013; see also Tan 2016). In sociologist Chua's (1994, 1997) highly influential work meditating on nostalgia for the Kampong, he refers to the affect as a symptom of the "politicization of stress" that is rooted in an imagined and simplified Kampong life. He describes how the fondly recalled slower pace of life in the village was in fact due to high unemployment rates, and the assumed safety a result of implicit public surveillance due to lack of privacy. In Chua's terms, nostalgia is an impotent sentiment, a critique of the present and a passive resistance to the relentless drive of modernization that has no real desire to engage with radical change or return to the past.

While I agree with Chua's findings to some extent, my thoughts equally consider Sophia and Sumit and their family. To them, their Kampong life is not a past to be exhibited, or a mere garnish for young couples' wedding photos. Nor is it like some of the heritage preservations that offer a sense of permanence. Rather it is situated in a thickened temporal realm, a home for their children, their elderly mother, and a place where they have willingly chosen to live, where they need to labor, appreciate, as well as respect their surroundings and the other species they share them with. What appears to render nostalgia and its associated way of remembering as a confusing and impotent sentiment is that some forms of nature and culture continue to be cast as a primitive past disconnected from the present. In other words, nostalgia loses its agency when it is understood only through the linear temporality of progress.

Noelani Goodyear-Ka'ōpua, in the context of indigenous Pacific activism, describes "expansive and sustainable futures" based on connections (2017, 191). She rejects a linear way of perceiving time that understands people protesting development as being stuck in the past. Rather, she argues that what is deemed to be the past within settler modernity ought instead to be understood as part of a more inclusive, multiplicitous futurity. Seen in this light, the act of countering the eroding effect of eco-modernization on futures and amnesia and disrupting the recursivity of double erasure cannot be one of merely holding onto things, or of preserving a static past that may further legitimize the logic of linear time and its "inherently progressivist, productionist and restless mode of futurity" (Puig de la Bellacasa 2015, 691). Instead, a more useful approach arises through an act of remembering that acknowledges the inextricability of the past, present, and future, and seeks to rebuild relations with others.[16] The act of remembering, pointing to the past and the future, must pay attention to multiple temporalities in the urban that, without ignoring the tensions in between, looks for alternative ways of imagining and relating. My participation at a local social enterprise group gave me a glimpse into these acts of restoring, repairing, and remembering.

It took hours to sweep the fallen leaves of the entire area. I swept until they formed a decent pile, then I put them into a wheelbarrow. When the barrow was full, I transferred them to the empty space adjacent to the in-house farms. Then I repeated the cycle. The leaves would later be used as natural fertilizer. There were approximately twenty of us for this weekend's Balik Kampung (meaning "going home" in the Malay language), a community program designed and organized by the Ground-Up Initiative (GUI), a nonprofit organization founded in 2008. Each of us focused on one task: clearing up the grass, preparing lunch, or sweeping the floors. The waste from our lunch of the day was separated into various compost bins. On the other side was a woodshed in which a few people were making bricks. Near the entrance of the hall, the small but bustling Saturday farmer's market was underway, and customers and farmers of GUI were greeting each other by names. The younger volunteers and I were delighted by the robust sunflowers. Seeing how we were amazed by the rich produce from GUI's own farm, a middle-aged farmer laughed saying, "you kids grew up with everything imported. Until the 1980s, Singapore was still self-sustained in terms of food."

I found GUI online. Its website link returned as the second Google search result when I used the keywords environment, Singapore, and Kampong. Its homepage features the following line: "Talent, is nurtured by the environment we live in." In 2017, GUI launched phase one of their vision, *Kampung Kampus*: A School of Life: a nature-inspired experiential learning space with an ambitious 26,000 sqm space in land-scarce Singapore. The founder, Tay Lai Hock (2012), highlights how the programs in GUI are designed to nurture grounded-ness and reconnect humans and the land. In his words (2012): "In Singapore, 'Land' is a property, something to be sold, and not something social. . . . People don't build their own houses anymore. When there is a disconnection between you, and the land . . . [w]e become apathetic and self centered. . . . We are not a green club, but a 'hands on' community; a community that wants to revive the good old kampung spirit. It isn't the kampung spirit that is archaic, stored in the museum."[17]

After finishing GUI's flagship program, Balik Kampung, I sat down with Chris,[18] one of the organizers of this largely volunteer-run group. I told Chris of my surprise on seeing the phrase Kampong spirit flooding the Singaporean media and wondered how GUI interpreted and applied it. Chris replied that the term Kampong spirit is loosely used these days, just like the word sustainability: "We do not want to romanticize the Kampong and its original meaning has been so diluted. To us, we prefer to use 'Kampong charm' . . . a 21st century embodied connection. . . . We are practicing it through the connection with environment and people, the 'ground up' way."

In a country where top-down planning is universal and "perfected," the name Ground-Up Initiative invokes a subtle resistance of its own. Although GUI and the government both emphasize the importance of engaging with natures in the urban context, each has approached it in their own way and arrived at different conclusions. The government endeavors to make the city "increasingly easy for residents to embrace a sustainable lifestyle that enables everyone to live comfortably while caring for the environment"

(MEWR and MND 2014, 22). For example, in order to encourage the community to visit the Sungei Buloh Wetland Reserve, an extensive mangrove boardwalk is advertised as bringing visitors "really close to the mangrove trees and wildlife without getting [their] feet wet or muddy" (NParks 2012). These mechanisms that promote a sustainable culture characterized by "convenience," "comfort," and "safety" without requiring residents to be close to nature or attend to environmental issues imbues the concept of livability and sustainability with disengagement and anthropocentrism. Comparatively, GUI believes that if "you've not been a direct participant in [the] green process, you're just an audience" (Tay 2015). In sum, as Chris explained to me, for GUI, leading a sustainable life in the urban environment starts with hands-on experience, taking more risks, remembering, and relearning the meaning of the land. In this way, it seems that remembering becomes an embodied and material act and effort to bring together some of the long-enforced distancing of intra and inter human-nature elements.

Guided by this approach, GUI has designed a range of social programs for participants of various ages and backgrounds, focusing on creating lively connection and reconnection with the environment: from growing vegetables, fruits, or herbs without pesticides to looking for useful ways to recycle and reuse unwanted materials for woodworking. According to many members of GUI, immense, painstaking efforts were put into the design and construction of the space. Most buildings were built by volunteers, brick by brick, and some certified for low carbon emission. I was also intrigued by some of the micro designs. When preparing lunch, water has to be fetched from the other side of the Kampus as there is no water tap in the kitchen. I later learned that this is one of the intentional inconveniences the GUI had incorporated in the design, in the hopes of encouraging the volunteers and participants of the programs to appreciate modern conveniences. This is particularly pertinent in a city that has long enjoyed efficiency and convenience. Creating a livable city, practicing sustainability in daily lives, and telling plural stories of a city calls for understanding and caring for its multispecies inhabitants. And care is never as easy as walking on a mangrove boardwalk. It is always "political, messy and dirty" (Puig de la Bellacasa 2015, 707).

Some volunteers at Balik Kampung mentioned that they appreciated the opportunity to focus on one task at a time, so I asked whether this was intentional. Chris smiled: "For today's program, you are asked to do one task. This is hoping you can first connect with the task on hand and connect to yourself. But GUI's practicing of slow and awareness is not just through doing one task a time like meditation . . . being slow is also to respect others, to allow them the necessary time to develop. . . . We are pragmatic idealists. There are lots of tensions and tasks to deal with. Like how can we pay the large sum of rent for this space next month?"

As the conversation continued, I learned that GUI was thinking of the possibility of expanding, to have a prawn farm and to grow rice. Chris told me that the members of the GUI were also still learning the complications and conflicts between the fast and the slow, between growing and sustaining, teaching and practicing. Chris's thoughts, and the practices and ethos of GUI, are enlightening. For it is crucial to understand that slowness is not necessarily a solution to counter the dominant discourse of productivity and pressing forward. Fastness, or speed, as Isabelle Stengers (2005) points out, could be an oversimplified position and a romanticized approach. Rather the art of practicing and inhabiting slowness resists reducing the complexity of time or being emancipated from a messy and entangled environment. In part disrupting the linear model of time and practicing the art of remembering is to engage/re-engage other actors to understand the tensions and relations at work. For example, Bastian's eco-temporality work challenges us to engage with more-than-human time, as a way to make visible our entanglement with the environment, in particular to understand the increasingly unstable time of an active Earth and "the frustratingly slow time of human efforts to respond to recognized environmental threats" (2012, 44).[19] In doing so, a more-than-human time complicates what is "often assumed to be a shared and all-encompassing present" and resists a universal time (Bastian, 45). From this perspective, sustainable and slow time in fact requires acceleration in order to respond more quickly to issues of environmental degradation.

In many ways, GUI's grounded approach to inhabiting time and the surroundings takes me back to Sophia and Sumit's home. Their way of being rooted and building relationships is grounded in local specificity and is present, rather than looking back to admire an imagined past. Of course, Sophia and Sumit have been grounded throughout their time in Kampong, which is embodied and labor-intensive rather than compressed or forcefully accelerated. This includes the "inconvenience" that they must cope with to live in and through the environment. Looking through this lens, to revive or recollect the Kampong spirit is not about holding onto a past, and certainly cannot be manufactured through the deployment of certain elements. Instead, it is enacted through an active way of remembering that cultivates both the ethos and the environment in which this spirit is embedded.

In a similar vein, GUI's program is designed to experiment with multiple rhythms, linking past and present through cultivating and caring for the environment and relations. Is it possible that this practice could counteract the double erasure and bring a "21st century Kampong culture" into being in a lively, rooted and imaginative way? When asked about his vision of GUI, Tay says: "hopefully GUI can convince the government, the people in this country, and hopefully the world to question, 'Should we really ask them to stop and relocate, or should we relook at our master plan? Did we miss something in the process of nation building, in the process of trying to plan everything to the tee" (2014, 100)? To GUI, the growth of their establishment, its Kampung Kampus and the attention it has received has come as a surprise. When interviewed for her movie *Moving House* and other work, the Singaporean director Tan Pin Pin said that she had seen the Singaporeans' "fighting spirit" emerge in the campaign to save Bukit Brown; perhaps there is still hope as she had not seen evidence of this spirit when her family had to move their ancestor's tomb (2012, 63).[20]

━━━━━━━━━━━━━━━━━━━

I opened this chapter with my visit to Sophia and Sumit, one of the few remaining Kampong residents on Singapore mainland, who resist exchanging their surroundings and the Kampong spirit for the four walls of an apartment and an imposed way of life. At the end of my exploration, I was in Kampung Kampus, with many others, learning how to practice re-grounding and cultivating the environment that nurtures us and may give rise to the Kampong spirit. In between, many humans and other-than-humans have helped me to navigate the multifarious process of urban development, from an island-wide public housing program to a highly calculated scheme for conservation and preservation, from the radically transformed offshore islands to the heritage trees that become the metonym of an imagined past, and to an eco-town that has rede-signed the sustainable narrative. At the same time, these are also stories of refusing a singular way of futuring, from the efforts to sustain the home of the living and the dead to cultivate an active remembering. Here, unlike monumentalizing a particular past (Hutton 1993), or chasing a singular version of future, to remember is to learn to live responsively and imagi-natively with the environment, and to respect one's limitations.

Importantly, this chapter examines modern and eco-modern amnesia, instigated and sustained by the intra-related erasures of the more-than-human environment and its entangled social, cultural, and ecological rela-tions. As the narrative of sustainability and development is increasingly reframed as a collaborative relationship, Singapore is seeking to establish itself as an embodiment of the reconciliation between the need of develop-ment and conservation (MEWR and MND 2014). However, as the necessity for, and impact of, social engineering and development plans are often assessed as isolated political or business cases based on a singular concep-tion of the future, the rich relational past is depleted, and the interdepen-dent socioecological connections are disrupted. The resulting sustainable fix becomes a combination of conserving an imagined past, sustaining the status quo with a repackaged eco-narrative while accelerating technologi-cal research of environmental solutions with an extremely narrow set of life possibilities.

This chapter shows some of the consequences and impacts of limiting and containing time and possible futures in urban development and the entangled reduction of cultural and biological diversity. Having reflected on the various threads I have woven together and the web of relations in which they are situated, I believe that there is a great urgency to challenge the normalizing process of relocation or uprooting within urbanization, to interrupt the momentum of world-making that is directed by linear devel-opmental time, and in so doing to open up to other temporal forms. In the era of amnesia, this chapter addresses the need to disrupt double erasure and intervene in displacement using a relational act of remembering. In this context, practicing remembering is an ethical and practical work that examines the absence, the things that are being canceled in the deep and recent past, and the rich relations that could have been forged. It asks us to cultivate a responsible approach that respects the relationships between memory, temporality, and the environment.

3

REIMAGINING URBAN MOVEMENT

"In Singapore, there are no traffic jams, the worst we experience is slow traffic." We started our tour of the LTA gallery with this statement from the delegate of the Singapore Land Transport Authority (LTA), a statutory board under the Ministry of Transport. I had come to the gallery to learn more about the history and future of Singapore's urban mobility. The audience consisted of about twenty people from various countries. Over the course of the next hour, we were introduced to a range of mechanisms used by the city-state to ensure that traffic moves at its optimal speed and maximum efficiency. These mechanisms include Electronic Road Pricing, a pay-as-you-drive system designed to manage road congestion, and hefty vehicle taxes to curb car ownership. As we were guided through the trajectory of Singapore's transport development, we were repeatedly reminded that transportation infrastructure has been the cornerstone of the island's nation building, and that its continuous expansion is crucial to further enhance connectivity and livability in the country.

At the conclusion of the visitors' program, we watched two animated videos featuring aspects of the island's future mobility planning. These stories, narrated from the perspective of a family of three generations, focused on a debate as to whether the construction of a new Mass Rapid Transit train line should be close to homes, which would bring noise and other issues, or away from homes, which would involve longer journeys for commuters. The family also had to decide between taking public transport or driving, each with its own advantages and disadvantages. After watching the videos, we too were invited to choose between the two sets of decisions on these issues by pressing the buttons installed on our chairs.

In this imagined, imminent future, the seemingly complicated challenges offered to participants did not mention how the construction of new train lines might disturb the natural environment or involve relocating existing inhabitants, both human and nonhuman. Transport infrastructure with its attendant issues of *whose* mobility is enabled, and how, is entangled in a range of ways with a cluster of social-cultural, political, economic, and environmental issues. Yet, in this admittedly simplified context, all these complexities were reduced to a simple set of yes or no questions centered on the trade-offs in terms of personal convenience and amenities. This experience of encountering a framing of urban mobility planning confined to a narrowed set of stakeholders and options impelled me to ask more questions.

In this chapter, I explore urban movement, paying specific attention to transportation. In Singapore's effort to build a sustainable and livable city, the framework for understanding livability comprises three core aspects: a competitive economy, a sustainable environment in which the city has to survive with limited space and natural resources, and a high quality of life (Centre for Liveable Cities and Civil Service College Singapore 2014). In line with this vision, the transport department outlines its goal to develop a people-centered land transport system where "all can get to more places faster and in greater comfort" (LTA 2013, 50). Yet, as geographer Tim Cresswell argues, creating models of mechanically aided physical movement to make transport more efficient, or less environmentally harmful, "says next to nothing about what these mobilities are made to mean or how they are practised" (2010, 19). Further, Cresswell points out the need to place "human mobilities in an entangled web of 'other' mobilities that have sometimes been demoted in mobilities research" (2014, 713).

Cresswell's concern for the "other" mobilities, namely animal movement, in an urban context has become an increasing concern to researchers. For example, cultural geographers Timothy Hodgetts and Jamie Lorimer (2018) explore how animals' mobilities are governed and shaped by human actions and their complex social and political implications. Philosopher Clare Palmer, in the context of urban environmental ethics, posits a causal and moral responsibility toward displaced wild animals in densely human-populated environments as animal territory is encroached on by development (2003, 71).

With these various urban mobilities in mind, moving alongside or against one another, my inquiry aims to tease out the complexity of mobility and its entanglement with technology, development, and urban natures. I focus on two seemingly contrasting yet intimately related case studies: a proposed cross-island underground train line earmarked to run beneath a nature reserve and an ecological bridge spanning a six-lane high-way that hopes to restore animal movement within the same reserve. In tracing the design of the train line, I am inspired by philosopher Paul Virilio, who is interested in how the discourses of urban mobility are directed by the desire for velocity as well as by the politics of (underground) invisibility and a fixation on certainty. As I navigate the wildlife overpass in the second section, I investigate the ways in which the technology of restoring animal mobility may open another way of world-making, and conversely, how it may further enable automobility, serving as an "ethical bypass" for particular development projects (Franklin 2001). Through the interplay of these two cases, I ask which forms of movement are promoted as desirable and which are cast into the background to become the "shadow places" that Val Plumwood (2008) describes as requiring greater attention.

In addressing these questions, this chapter unsettles some taken-for-granted approaches to urban transportation planning, which is underpinned by a measurable, singular imagining of the future. Thinking with and responding to the intersection of work on transportation, urban nature, and multispecies mobilities, it offers a new mode of inquiry into urban movement by exploring the development of supposedly sustainable yet largely human-centric infrastructure projects, along with emergent social, economic, political, and environmental issues espoused by various types of movements. Amid the growing expansion of infrastructure and public transportation in Singapore and around the world, often in the name of sustainability and livability, examining urban movement through a more-than-human lens is essential and urgent ethical work that may open up spaces to rethink possibilities for a more diverse and flourishing multispecies movement.

INTERCONNECTED MOBILITIES: A GLIMPSE INTO THE PAST

Although Singapore was not heavily transformed by its own desires for internal mobility until its independence in 1965, distant mobilities and others' aspirations were central to its shaping much earlier. Originally established as a trading post for the British East India Company in 1819, the island soon became an international port. The bustling trade movement would later cause severe damage to its coastal environment. Meanwhile, its landscape was also completely transformed. In less than a century of British colonial rule, almost all the primary lowland tropical rainforests of the island were cleared to make way for plantations (Castelletta, Thiollay, and Sodhi 2005). Initially, the plantations consisted largely of gambier used in the British colonial dyeing and tanning industry, but by the end of the nineteenth century, fueled by the invention of pneumatic tires, Singapore became central to the rubber growing industry (Thulaja n.d.). This transformation was needed to meet the rapidly increasing demand for automobiles that was at this time transforming mobility in other parts of the world (O'Dempsey 2014).

Since its independence, Singapore has joined the forces of global capitalism and modernization, shifting from a reluctant enabler of others' mobilities to a determined proponent of frictionless movement. In Singapore, transportation has developed into something more than its apparent function of meeting travel demands. It has become "a nation-building project aimed at realizing the state's vision of a business-friendly, smoothly flowing urban economic unit" (Lin 2012, 2482). A dense network of six-lane arterial roads was created, with semi-expressways equipped with flyovers or underpasses to further reduce traffic delays (Barter 2008, 103). In 1987, the first Mass Rapid Transit train line was opened. Its speed and ability to maintain a less disrupted service soon positioned it as the backbone of Singapore's transport.

To maintain its forward momentum in one of the most densely populated countries in the world, the state has placed a high premium on enhancing its transportation network, often at considerable cost to other kinds of land use and the environment. A program of island-wide urbanization along with the aggressive expansion of transport infrastructure resulted in both deforestation and the demolition of most Kampongs and heritage sites. As discussed in the previous chapter, the highways and traffic tunnels that were constructed, "devoid of meaning and history" (Leow 2010, 126), replaced the beloved national library, national theater, and cemeteries that carried generations of local people's memories.

This brief moment into the past suggests some of the ways in which mobility has shaped this island-nation and the futures it imagines and pursues. In a discussion of urban planning, it is important to remember that mobility and immobility, including each of their impacts and consequences, are never singular but interrelated events. As we will see, as the young country of Singapore tries to negotiate pathways out of its past, an ethos of velocity and economic progress has arisen that to date continues to underpin its policies in urban development. As a result, certain kinds of movements that are enlisted to press forward with this development regime tend to be given preference over both other movements and efforts to stay in place.

DEBATING THE CROSS ISLAND LINE

In 2013, the LTA published the updated Land Transport Master Plan (Masterplan thereafter), in which the construction of the Cross Island Line (CRL) was proposed as the eighth Mass Rapid Transit (MRT) rail line in Singapore. This line is set to play a key role as part of the country's livability goal. Targeted for completion by 2030, the CRL is to run from the suburb of Changi to Jurong, connecting the east to the west.

Besides relieving the load on existing MRT lines, the CRL claims to provide commuters with greater comfort and shorter journey times. Since the proposal was first released, the CRL has caused ongoing heated debate, in part because it involves going underneath Singapore's largest remaining reserve: the Central Catchment Nature Reserve (Central Reserve). The Central Reserve, which occupies over two thousand hectares, has some of the country's richest forests in terms of biodiversity and is home to most of its very few remaining pristine freshwater streams (NParks 2018).

In light of the island-nation's already precarious ecological situation as a result of its long privileging of development, including transportation

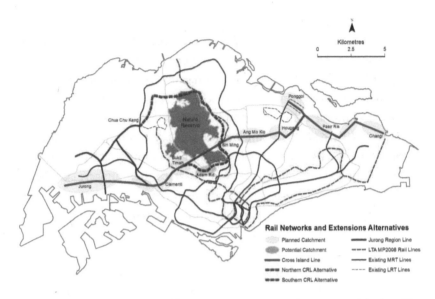

3.1 Alternative routes proposed by NSS, 2013. Image credit to National Society (Singapore). The original map is included as figure 16 in "Cross Island Line: Discussion and Position Paper" (NSS 2013).

networks, it becomes clear why the proposed rail link is being challenged by environmental groups. One outspoken critic of the proposal is the Nature Society (Singapore) (NSS thereafter). The NSS has highlighted the fact that the Central Reserve has already been broken up by "reservoirs, pipelines, sealed roads, military facilities and security fences" and that any further fragmentation and disturbance would have a serious detrimental effect on its ecological condition (NSS 2013, 11). It further proposed two alternative routes for the CRL that run around the reserve (see figure 3.1), of which the southern one may even allow the train line to serve more residents and commuters in that vicinity.[1] Some local Singaporean residents felt equally strongly about the potential environmental impacts and organized campaigns to resist the building of the CRL. The Love MacRitchie Movement—an effort among various nature groups in Singapore—was spontaneously formed in 2013 in response to the CRL. The group went on to organize a series of walks, tours, and campaigns to introduce the rich and delicate ecosystem of the nature reserve and raise the awareness of potential conservation issues surrounding the proposed train line.[2]

Following extended discussions with environmental groups, the transport authority announced that they would now consider two route options for the CRL: a direct four-kilometer alignment with two kilometers of tunnel running underneath the Central Reserve, and an alternative nine-kilometer route that skirts the reserve boundary in which sections of the line may go underneath existing homes and businesses (Singapore Government 2016). The authority warned that the skirting option could increase the construction cost by two billion dollars and cost commuters six minutes of additional travel time. In this context, they noted that, "because in the mindset of the MRT commuters, an extra half a minute is already terrible" (Hansard 2016).[3] The final alignment of the CRL would not be decided on until 2019, six years after the construction plan of the underground train line was first announced.

In a dynamic and compressed urban environment, the preferred form of mobility is often highly political and negotiated between contested forces. For example, due to land constraints in Singapore, space-efficient and affordable mass mobility is said to take priority while car ownership is strongly discouraged through hefty taxes. And yet, transport researcher Paul Barter points out that not only do cars in Singapore enjoy a large speed advantage over public transport due to road designs, but this situation seems to have inspired a greater social acceptance and a strong sense of elitism in relation to car ownership (2008, 101). Despite these difficulties, public transport is central to Singapore's efforts to stimulate economic growth and increase population, as well as to catch up and become a leader in sustainable urban mobility. The Masterplan states that by 2030 Singapore should double its rail network, with 80 percent of households being within a ten-minute walk from a train station (LTA 2013, 2019). In doing so, it is hoped that the nation will be able to increase its capacity in mass transit while reducing its reliance on cars and carbon emissions in the transport sector. The proposed CRL is a critical component of this vision. So, how has a seemingly more sustainable form of transport sparked heated debates over its environmental consequences? Why have the LTA continued to insist on considering both routes, despite the fact that the skirting option would circumvent the reserve as well as serve more residents as the NSS and a member of the Parliament have argued?[4]

VELOCITY: THE UNBEARABLE SIX MINUTES

There is an inextricable connection between mobility, velocity, time, and development at work in these debates over the CRL. French urbanist Paul Virilio (2001, 2006) proposes that dromology, defined as the "science or the logic of speed," is now a central force shaping social, political, and cultural development in the modern and postmodern world. For Virilio, Verena Andermatt Conley writes, "the Cold War, whose end signified not so much the triumph of liberalism as the failure of a type of social experimentation, has been replaced by a global economic war based on speed" (2012, 86). Indeed, both modernization and mobility share an insatiable appetite for harnessing velocity because of its ability to compress distance, space, and time. Singapore sociologist Chua (2011) maintains that modern Singapore's "success as identity" is built on its hyper-accelerated miraculous transformation from third world to first, made possible by its transportation infrastructure, high technology, and the intense flow of capital (see also Chong 2010). As the ability to move fast is tied to economic competitiveness, the narrative of urban mobility is underscored by speed, efficiency, and persistent growth.

In discussing the interrelationship between capital, space, speed, and time, Jason Moore, quoting Marx, notes that "capital incessantly drives towards the 'annihilation of space by time'" (2015, 10).[5] He further writes that "Capital seeks to create a world in which the speed of capital flows—its turnover time—constantly accelerates. The privileging of time over space in capital's project is not passive but active: every effort to accelerate turnover time implies a simultaneous restructuring of space" (10). As Singapore fully acknowledges its limited natural and human resources and intense space-scarcity, the focus of its urban development has persistently been on the optimization of its land and sea space. It seems that the more the city-state centers on its spatial limitations, the more it attempts to compensate for or overcome these with efficient transportation and communication networks. Locally, the expansion of the train network allows it to compress space and further the progression of urbanization. Internationally, it allows the island-state to establish a social-economic model with a global hinterland. The Port of Singapore is, after all, one of the busiest in the world in terms of shipping tonnage. It is in this context that the CRL must be understood. As velocity continues to drive development, the additional six minutes arising from the proposed skirting alignment of the CRL is perceived as an unbearable disturbance to commuters, and a disruption to the forward momentum of transport and economic development. This is the logic that has become central to the state's position in defense of the direct route.

Furthermore, it is important to note that although they are greatly emphasized, velocity and efficiency are not evenly distributed. In other words, mobility is always hierarchical. In his work on the politics of mobility, Cresswell maintains that "Being able to get somewhere quickly is increasingly associated with exclusivity" (2010, 23). This is certainly the case in Singapore where, as Barter points out, due to the high cost of car ownership, there is constant pressure on the state to provide "a high level of service for private motorised traffic" (2013, 237). Mobilities researcher Weiqiang Lin argues that as cars enjoy a large speed advantage over public transport, one of the greatest inefficiencies in traveling by public transport in Singapore is "the disproportionate amount of time required to traverse relatively short distances" (2012, 2487). Furthermore, public transport is, in fact, entirely managed by private companies, an arrangement implying the least cost to a state that subscribes to "the logic of market allocation" (Lin 2012, 2486). Indeed, during the tour at the LTA, we were informed that a government agency is not the service provider for trains or buses, but rather the regulator who monitors their service standards. Under these privatized arrangements, Lin argues that travel time has been "commodified" and "monetized" through a set of "strategic pricing mechanisms" (2484). Thus, "the temporal dimension of urban travel is differentiated and enhanced for some at the expense of others" (2481).

In 2011, the deteriorating condition of public transport services caused rare open public discontent, culminating in the early "retirement" of the minister of transport (Lin 2012, 2477). Given the conjunction of disappointing public bus services and an insistence on prioritizing road usage for cars, the state has put more and more emphasis on expanding the underground rail link as perhaps the only viable alternative mode of transportation "that can even remotely compete with the car on speed for long urban trips" (Barter 2013, 237). When commenting on the route options of the CRL, the transport minister, Khaw Boo Wan, said that even an extra one minute of travel time is a lot of time, let alone six minutes. In his words: "When the train has a disruption, causing an extra one minute of delay, commuters can, within that one minute, send off, maybe, 100 tweets to flame LTA or SMRT" (Hansard 2016). The state's anxiety over speed is highlighted here, but so too is its own contribution to the constant push for efficiency and speed. In this light, this tension over an extra few minutes of travel time seems to point less to the public's thirst for speed than it does to a growing resistance to the inherent inequality in Singapore's urban mobility planning and the underlying social injustice. It is as a result of these interwoven factors that shortening commute times becomes more a political necessity for the ruling party than merely a question of meeting travel demands.

UNDERGROUND: THE POLITICS OF INVISIBILITY

In his research on Singapore's urban mobility planning, Barter suggests that the country's primary effort "to improve the alternatives to cars has been rapid expansion of the rail system" (2013, 235). It is "a 'World City'-inspired strategy, looking to London, New York, Tokyo and Paris where large percentages of people, including affluent individuals, rely on urban rail for mobility" (Barter, 235). In contrast, the bus system, while remaining important, "has been relatively neglected" (235). In the Masterplan, walking and cycling are briefly mentioned. However, as they are positioned primarily as a means of transit from the train or bus station to home, this possibly further encourages the construction of more MRT stations throughout the small island.

In outlining the two possible alignments of the CRL, the transport authority stresses that although the direct route cuts through the Central Reserve, it will do so underground with a two-kilometer tunnel that runs forty meters below the surface.[6] There will be no construction at the surface level. In contrast, the transport authority notes that because the skirting option requires longer tunnels, it will need extra "ugly" ventilation shafts to be constructed at the surface (Hansard 2016). Clearly, the politics of visibility and invisibility is at work here, which is to say, things are placed underground and thus out of sight as part of an effort to resolve the problem. Plumwood highlights the danger of the neglected, denied and "unrecognised, shadow places that provide our material and ecological support" (2008, 139). Although Plumwood's focus is on a different aspect of environmental justice, it is useful to think with this notion to make visible some of the ethical and justice issues relating to the underground.

In urban expansion, a large part of what enables the further development of the underground train network—rather than improving bus services or taking diversification of urban mobility seriously—is the seeming "invisibility" of the subsurface. Yet, discussions and questions often have to go much deeper into these shadow places. Recent research has explored how we have imagined and disenchanted the subterranean in ways that might reduce it to a mere resource in urban development or a burial site for toxic and radioactive waste (Kearnes and Rickards 2017). What are some of the associated risks of underground activities? In their work, Maria de Lourdes Melo Zurita, Paul Munro, and Donna Houston argue that the emergence of capitalization and industrialization "are predicated on new technologies that redefine the underground as an epistemological space for economic, social and political calculation" (2018, 298). Indeed, the subterranean is not a smooth space, and things are often more complex and interconnected than is immediately apparent.

Although not always readily visible, geologists, engineers, and environmental researchers have identified many problems with underground works including noise, soil erosion, salt pollution, loss of biodiversity, and pollution of groundwater and air (Gattinoni, Pizzarotti, and Scesi 2014). In the case of CRL, *underground* seems to create an impression that there would be no significant damage to the Central Reserve. However, many of these issues need to be considered. The NSS, in particular, have highlighted the fact that mandatory geotechnical site investigation of the project would involve intensive borehole drilling operations within the reserve and extensive clearance of vegetation to prepare the sites (NSS 2013, 3). During my interview with Tony O'Dempsey, a conservationist who played a key role in negotiating the CRL issue with the officials, he explained to me that these site tests produce their own set of problems and challenges: "I am most concerned about those boreholes. The ones that are needed for the engineering works on the surface for soil investigation. We are worried about the silt and other types of pollutants that will be released into the streams as a result of the soil test, which is likely to be the greatest direct threat to the stream ecosystems of the reserve. It will permanently destroy some of the most pristine stream habitats remaining in Singapore."[7]

In 2016, the first phase of the Environmental Impact Assessment (EIA) on Central Catchment Nature Reserve in regard to soil investigation for the CRL was released. Echoing some of the concerns raised by the NSS, it pointed out that aspects of site investigation activities could lead to disturbance to vegetation and wildlife behavior, disturbance of aquatic habitats, and more (Environmental Resources Management 2016). In addition, it highlighted the possibility of unplanned events that could cause a range of ecological issues such as unplanned water pollution from site runoff, injury to wildlife including roadkill, and habitat loss. The EIA concluded that, from an ecological and biodiversity perspective, the impact on the reserve will only be kept to "moderate" levels even if comprehensive mitigating measures are effectively carried out. In contrast, the impact from the skirting alignment is negligible—at least with regard to the Central Reserve.

In the same year, following extended engagement with conservation groups, the transport authority announced that they would reduce the number of boreholes for the site investigation of the CRL project from seventy-two to sixteen, to be placed within the existing public trails to minimize disturbance to vegetation (LTA 2016). It promised that more mitigation measures would be introduced to control the environmental impact. In early 2017, the soil investigation started. Tony told me that not everyone was happy with this reduced scope of soil testing: "The soil investigation so far is going well. But some of the nature communities do not think we have won as the train may still go underneath the reserve. I think we have been successful this time compared to some of the other conservation battles we fought." Tony went on to tell me his concerns about the accumulations of the insidious and damaging effects of continuous fragmentation of forests and green spaces in Singapore and the compromises that were required to enable ongoing conservation efforts.

In addition to hosting an extensive transportation network and shopping malls, Singapore's underground has been used as a storage space for petroleum and ammunition. In these efforts, various others have been displaced. As discussed in chapter 2, since 1978, the government has had the power to acquire all private cemeteries when required. Since this time, thousands of graves have been exhumed to make way for transport infrastructure or public housing. If some parts of the underground were once imagined as the dwelling site for the dead as well as a liminal zone for the living to connect to the past through tomb-sweeping or other rituals, years of exhumation of the ancestors' tombs and the resultant, gradual erosion of associated memories and cultures has relegated the underground to merely "empty" usable spaces.

In 2015, amendments to the legislation were made to expand the state's land acquisition power to underground and air space (Ministry of Law 2015; Chew 2017). Soon after, as part of Draft Master Plan 2019 for an Inclusive, Sustainable and Resilient City, a three-dimensional underground plan that maps out three districts of Singapore's subterranean spaces was unveiled. According to the Urban Redevelopment Authority (2019), the aim is to "free up surface land for people-centric uses by relocating utilities, transport, storage and industrial facilities underground" for further future growth. For Melo Zurita et al., it seems that as the underground spaces are "harder to see, less accessible and directly experienced . . . they are perceived as less charismatic" and as "more pollutable" (2018, 302).[8] As subterranean spaces grow increasingly tantalizing for urban development, it is all the more urgent to ask, "What happens in places that are out of mind, out of sight?" (302).

CERTAINTY: THE CONTROLLABLE ENVIRONMENTAL IMPACTS

Since its independence, Singapore's accelerated modernization and robust economic development through meticulous planning, controlling, and engineering human and nonhuman elements has been the cornerstone for the PAP. Its ability to plan to perfection has been held up as a model globally and packaged and exported to other countries.[9] Of course, the future mobility of the city-state is being planned with the same assurance. In the Masterplan, the state identifies an increased "demand for transport" driven by a greater population: "Population growth in the last five years has brought new urgency to some of the plans we had already outlined. Our economy is also expected to continue to grow at an average annual rate of 3 to 4 percent over the next 10 to 15 years so we can expect that more people and goods will need to be transported. We are already planning for infrastructure to meet this future demand as this will take 10 years or more from conceptualisation to completion" (LTA 2013, 12–13).

Perhaps unsurprisingly, the proposal for the new train line was released alongside the now infamous Population White Paper (National Population and Talent Division 2013), which proclaims that by 2030 the population of Singapore needs to reach 6.5 to 6.9 million people (from 5.3 million as of 2013) to maintain its economic outlook and development momentum. With one of the world's lowest fertility rates, this projected population growth requires a further increase of carefully selected skilled immigrants. As discussed in chapter 1, Singapore adopts distinct policies governing foreign workers classified as less skilled or as foreign talent. In a careful process of creating a sustainable and livable city, the city-state strictly outlines whose contributions are included and whose are to be placed in the shadows. Facing a rare protest from its people, the government insisted that the White Paper "was not about any specific population size for beyond 2020, but rather that it was being used for the purpose of land use and infrastructure planning" (Yahoo 2013). As the need for mobility is dictated by a top-down policy grounded in a firm belief that the future must be systematically planned and contained, the potential sacrifice of, and damage to, the Central Reserve is based on a manufactured and quantifiable demand for future mobility, envisioned by a single ruling party fixated on sustained economic growth, technological advancement, and a supervised population mix.

In many ways, the appeal of framing things with certainty lies in a seemingly simplified account of issues, which in turn creates the illusion that things are always controllable. In his work on statecraft, James C. Scott discussed extensively this dynamic of control and manipulation through scientific rationality and schematic knowledge brought about by focusing on "certain limited aspects of an otherwise far more complex and unwieldy reality" (1998, 11). In response to public inquiries about the potential impacts of the direct alignment of the reserve, the LTA shifted focus to their certainty of mitigating and containing the risks. In doing so, the tension of cutting across a highly ecologically sensitive area and the related complex environmental, social, and cultural issues are reduced to a controllable set of measures of engineering and science. Yet, complete reliance on mitigation measures is not only problematic but also unrealistic. In calling for zero impact on the CRL, various "green groups" expressed their concerns to the media (Tan 2016). For instance, biologist David Tan from Love MacRitchie

Movement highlighted an earlier incident that a stream in a forest just outside the Central Reserve was polluted by contractors despite mitigation measures being in place: "Similar lapses could affect the rich biodiversity in the reserve."; Natalia Huang, an ecologist, cast doubts on the efficacy of mitigation, noting that "[H]ow much of the impact is mitigated cannot be guaranteed and may not be measured either" (see Tan).[10] It seems that when the state considers the alignment options and associated potential damage to the reserve, it expresses much certainty and places emphasis on its ability to mitigate and to decide what is best for all. However, it shows little respect for speculative conservation issues.

In recent years, researchers in urban ecology and more-than-human studies have been challenging this dominance concerning the certainty of knowing (the known). Instead, they assert the importance of "understanding the limitations of that knowing, its uncertainties and indeterminacies" (Hinchliffe and Lavau 2013, 271; see Tsing 2015). As Steve Hinchliffe and Sarah Whatmore explain, a living city consists of experimental activities with unavoidable "elements of not knowing, the unknown and the unknowable" (2006, 130). In this light, considerations of urban mobility, including its associated risks, cannot stop at the edge of what we know or think we may know, and are therefore able to control or mitigate. Rather this situation demands that we search more widely and respect what we may not know, and may never know.

As Singapore plans its future, environmental issues have not only been factored into its planning but also said to offer substantial opportunities for green growth through job creation and the exporting of low-carbon technological solutions (MEWR and MND 2016a). As cities are responsible for more than 70 percent of global carbon dioxide emissions, increased attention has been paid to the need for cities to transition to a low-carbon future (Bulkeley et al. 2011). Singapore's Climate Action Plan covering the period of 2016 to 2030 states that in the process of building a carbon-efficient nation, the country seeks to reduce energy costs while bringing economic benefits, and there "may also be co-benefits such as a cleaner environment" (MEWR and MND, 50). Indeed, much of the emphasis on trains as having a positive environmental outcome is because sustainable mobility is often narrowed down to an issue of quantifiable and measurable CO_2 reduction benefits while all other environmental impacts are framed as necessary costs, or as risks that can be controlled through scientific measures. As a result, the dynamics of sustainability are dangerously reduced and run the risk of being recruited into development regimes.

Interestingly, in assessing the dramatic increase in rail lines in China, researchers argue that the expansion of urban rail transit will not reduce the total number of cars as it would "promote population increase and economic development, and further stimulate car ownership" (Yang, Zhang, and Ni 2014, 530). While the inhabitants of Singapore, human and nonhuman are awaiting the decision on the final CRL alignment pending the phase two release of the EIA, it has become clear that the issue of urban mobility is much more complicated than advocating for a narrower version of sustainable transport or a battle between green groups and the state, as many media outlets tend to portray it. Rather it is a process entangled with the desire for velocity, enabling a politics of invisibility and the alluring perception of being in control. At the same time, in contrast to the underground rail project that threatens to further fragment the Central Reserve, another, much more visible, repair work is taking place on the edge of the same reserve.

A CELEBRATED BRIDGE

There were no signs indicating the direction of the bridge. Often noises from the highway were my guide. So as not to trespass on the restricted area or disturb the animals that may be searching for a way to the bridge, I decided to move toward the edge of the reserve where it met the road. The path I took was perhaps not often trodden. Bushes were overgrown, impenetrable. Tree branches were thick, tightly woven, and entangled, leaving almost no gap where I could walk. There were times when I had to bend low and really close to the ground. As I tried to remember the skills of moving on all fours, my body began to assume the steps of a pangolin or a civet, moving carefully but also somewhat aimlessly, hoping that a green field may appear soon to reconnect me to the forest on the other side of the highway.

After quite some walking and crawling, I arrived at the border of the forest adjacent to a six-lane highway, where the thick green was separated from grey concrete only by a narrow ditch. There, I had an unobstructed

3.2 Eco-Link@BKE, view from the edge of the reserve. *Source:* photo by author.

front view of the bridge I was looking for with its large signage decorated with animal figures, "Eco-Link@BKE" (see figure 3.2). The bridge was still in its first few years of use, and the vegetation growing on it was thinly formed. I had seen satellite and aerial images of it on my computer. I had passed underneath it numerous times as a passenger in a moving vehicle. Yet, in contrast to these distant encounters, seeing it in this manner hit me in a visceral way. Without being shielded by the steel structure of a car, my body was intensely aware of how the whole environment trembled as it endured the velocity of nonstop traffic, including heavy trucks moving along the expressway. The noise and vibrations were almost unbearable, making me anxious and uncomfortable.

The green area I walked through on my way to the bridge is part of the Central Reserve, the same forest that is at the center of the CRL controversy. Even without the proposed train line, the reserve is already a thoroughly contested landscape. In 1986, the eleven-kilometer six-lane Bukit Timah Expressway (BKE) was constructed, separating the Central Reserve and its adjacent Bukit Timah Nature Reserve. Since then, the highway that supposedly saves motorists thirty minutes in travel time has cast a long shadow on the ecological conditions of the once continuous forest, from the reduction of flora and fauna to the rise in roadkill. After disrupting animal mobility on the two reserves for decades, in 2013, a 62-meter-long ecological bridge, Eco-Link@BKE (eco-link), was completed. The bridge is not for human use. Rather, it hopes to restore an ecological connection by allowing animals to move freely between the two forests, thus expanding their habitat and linking genetic populations in an attempt to increase their species' survival chances (Fei and Li 2014). One prominent feature of this ecological path is its wide hourglass shape measuring fifty meters at the narrowest point. This width was determined by local conservation experts as the minimum width for some animals—including Sunda pangolins, Common Palm civets, and small native mammals such as squirrels or shrub birds—to "feel comfortable" when crossing the bridge (Chew and Pazos 2015).[11]

In more recent years, amid ferocious appetite for highway expansion, intensified urbanization, and concomitant habitat fragmentation, development of wildlife corridors is increasingly being taken up around the world and in the urban area (Clark, Zolnikov, and Furio 2021; Abrahms et al. 2017). The construction of probably the world's largest urban wildlife crossing is currently underway near Los Angeles, the United States, aiming to create a safe passage for a range of animals including mountain lions (Witt 2022). In addition to enhancing animal conservations and human safety, this connectivity technology has also been advocated for economic reasons as each vehicle-animal collision is said to result in significant financial cost (Huijser et al. 2009).[12] In fact, the concept of wildlife crossings is not entirely new. The world's first animal crossing was said to be constructed in France in the 1950s, soon followed by Germany and the Netherlands (Tepper 2011).[13] This practice was later taken up in various National Parks in North America (Goldfarb 2021).

Wildlife corridors come in a variety of forms, from the more commonly known overpass bridges to underpasses and viaducts.[14] In 2019, the first road canopy bridge in Malaysia was built to assist arboreal wildlife to move between fragmented habitats (Yap et al. 2022). At the same time, the data on the efficacy of wildlife corridors—from the extent of their restoring ecological connectivity and effects on "long-term persistence of [wildlife] populations"—remains limited (Ogden 2015, 452; Haddad et al. 2014).[15]

Some researchers suggest that non-biologists can easily connect with the concept (Jain et al. 2014; see also Van Der Windt and Swart 2008).[16] Others argue that animal movements are not always welcome. For example, invasive species might become unintended users of the corridors to travel and expand their range (Low 2013). Although researchers hold diverse and sometimes conflicting views in regard to these green passages, there is some consensus that the minimum width of wildlife overpasses generally needs to be over fifty meters to be ecologically sound, and the design of corridors needs to be specific, tailored to the animal users and the site (Brennan et al. 2022).[17] There is no universal model or script that can simply be replicated and mobilized as a quick solution.

Although discussions often focus on the structure of the corridors or scientific evidence of their use by target animals, in her analysis, Alexandra Koelle (2012) turns to the soft tissues of this connectivity technology. When she joined road ecologists collecting data on road-killed western painted turtles in the United States, she was acutely aware of the speed and volume of cars and even the danger of researchers themselves becoming roadkill. Koelle highlights the embodied care that biologists and road ecologists have practiced and their attention "to the movements, habits, and preferences of animals" in evaluating wildlife crossing structures (662).[18]

In Singapore, substantial efforts have also been put into the design and construction of the eco-link, from the structure of the bridge to the composition of the plants as well as the source of the soil. Years before construction commenced, the NParks embarked on surveys of the area including installing camera traps to monitor animals (Fei, Wah, and Chan 2016). In doing so, they helped the researchers understand and "guess" what animals may want from the bridge (Chew and Pazos 2015). Since the eco-link came into use in 2013, night cameras have captured the images of diverse users including pangolins, slender squirrels, civets, and various species of birds and snakes. As the trees grow taller, those using the bridge are expected to vary even further with the change in conditions. The eco-link, composed of a dynamic composition of materials, will continue to evolve through the intertwined rhythms of vegetation and animals. Indeed, the assemblage of this structural linkage and its users inspires a more lively and multi-relational approach to rethinking mobility through "a politics of conviviality" as proposed by Hinchliffe and Whatmore, which takes seriously "the heterogeneous company and messy business of living together" (2006, 134). From this perspective, in a time that some supposedly sustainable transportation may continue to interrupt the animal mobility, the bridge performs an alternative and experimental way in which urban nature may be imagined, resisting a singular way of moving.

A FUNCTIONAL AND ETHICAL BYPASS

As I continued my exploration, I found that not everyone was thrilled by the wildlife bridge. During my visit to Animal Concerns Research and Education Society (ACRES) wildlife rescue center, I learned of the average number of phone calls they received for animal assistance including roadkill. This turned the conversation to their view on the eco-link bridge. The staff member at ACRES told me that there had been frequent roadkill since the construction of the BKE highway as animals continued to move between the two very important yet separated forests. And now, although some animals may have found the eco-link, the roadkill continues outside the immediate vicinity of the bridge as the barricades along the forest consist only of bushes, which do not stop animals running through and onto the road. After all, as they noted, "Animals wouldn't know, *hey I shouldn't cross here, there is an eco-link coming up in 40 meters.*"

The frustration and scepticism of groups like ACRES is understandable. I was unable to find any record of a fence installed along the edge of the reserve, and certainly did not encounter one on my walks. Akin to maps for humans, fences are often thought to be a necessary part of green passages to "prevent the animals from entering the roadway, and [to] funnel them to the crossing structures" (Koelle 2012, 659; see Clevenger and Husijer 2011). At the same time, although the bridge may not be obvious to some animals, it seems to be highly visible to humans. In a *Mongabay's* article focusing on wildlife corridors, Stephen Caffyn, the director of the land architecture firm that assisted in devising the eco-link, points out that some of their favorite ideas for the design of the eco-link "stayed on the drawing board," including funnel fencing (Chen 2017). Caffyn goes on to explain that they always intended to include humans as a possible group of end-users: "The signage and façade of the bridge is something that we thought would be a simple and effective way to spread the word about the function of the bridge to motorists passing below and gives it an identity that is instantly recognisable even when traveling at speed" (see Chen).[19]

In reality, this particular "end-use" of the bridge has traveled much further than the designer possibly had in mind. The eco-link, alongside other green initiatives—such as the Park Connector Network, a green matrix of paths connecting parks and nature areas—has been heavily promoted as part of Singapore's environmental planning and its commitment to preserving biodiversity in the urban landscape. It is viewed as integral to pursuing the cachet of the world's leading sustainable city. Yet, despite the strong rhetoric directed at the public concerning these park connectors, some ecologists have questioned their ecological value for wildlife movement, as most of these green paths are only "a few meters wide" and consist of "pedestrian or cycling trails with sparse trees and shrubs," and see them more as a way for human connection in a more "natural" pleasant way (Jain et al. 2014, 889).[20]

Koelle encountered a related situation in her study in the United States, where although wildlife underpasses for some animals can be used more effectively and cheaply, it is the highly visible overpass bridges that continue to draw attention and attract funding, despite the uncertainties surrounding their value (2012, 660). As urban researchers Ian Douglas and Jonathan Sadler put it, the popularization of greenway in urban planning around the world may perhaps have established a "mythology of its own" (2011, 280). Indeed, recent analyses suggest that "planners, scientists and civil society actors need to be realistic" about the potential and limitations of green infrastructure, for example, wildlife corridors in terms of biodiversity and habitat restoration (Garmendia et al. 2016, 317).

The more I reflect on the overtly visible ecological overpass and the proposed underground CRL rail link, the clearer it becomes that it is this strategy of (in)visibility that has successfully distracted us from the irony of this situation. That is, the reserve that the proposed train line threatens to cut through (even if underground) is in fact the same fragmented habitat that the eco-link attempts to reconnect. As I recall the nonstop traffic passing under the eco-link, I begin to realize that the heavily publicized bridge is not only an ecological path to encourage animal movement but also, from the perspective of the state, an avenue for its continued forward momentum in the development of transformative infrastructure involving minimal friction. In this way, the eco-link and similar projects are used as functional technology and an "ethical bypass" (Franklin 2001). This is to say that whatever it might do for wildlife, the bridge also enables the continuity and expansion of auto-mobility both in its actual "working" and in its giving the appearance of having "fixed" the environmental issue associated with a highway running through the Central Reserve.

As historian Gary Kroll observes, a primary role of projects such as the eco-link is to counter the "conflicts between animal mobility and automobility" that results in roadkill; "a problem for drivers, [and] a challenge to the highway engineer's goal of creating fast and friction-free mobility" (2015, 4 and 15). Although the issue of roadkill has not attracted as much attention in Singapore as in some other countries, the deaths of the globally endangered pangolin while trying to cross the highway were highlighted as a key motivation for building the bridge (Chew and Pazos 2015).[21] Other commonly reported casualties are macaques, wild boars, leopards, and various reptiles and amphibians. The recent vehicular death of a rare sambar deer, of which fewer than twenty remain in Singapore, has given rise to a call to reduce the speed limit on one particular road. It has also sparked questions about increasing animal movement in the Mandai area and its connection to recent development projects (discussed next). As Barter notes, since congestion has often been equated with economic paralysis in Singapore, the authority is reluctant to lower speed limits (2013, 234).[22] Thus, the bridge that may reduce the number of roadkills becomes a functional structure to minimize potential disruption, thereby ensuring free-flowing traffic that supports the velocity the city prioritizes.

Furthermore, the eco-link and similar environmental measures sometimes play an ambivalent role in conservation, where they are subject to being instrumentalized to legitimize further disruption in other areas. Sarah Franklin (2001) proposes the concept of an ethical bypass that allows us to design around culturally and ethically difficult questions and issues.[23] In my discussion with a Singaporean conservationist, he alerted me to a recent large-scale eco-tourism project: "The development of the Mandai Safari Theme Park has started. It is said to be an eco-tourist mega-attraction that will combine the existing zoo, a bird park to be relocated from its current Jurong site, and a rainforest park. The existing secondary forest needs to be cleared up to make space for these new eco sites. The solution for dispersed animals is to follow the example of Eco-Link@BKE, and to build an eco-link@Mandai."

The area of Mandai, the site of a new eco-tourism hub in the making, has a rich and diverse population of wildlife as it sits on a secondary forest adjacent to the Central Reserve (NSS 2016).[24] Although a width of fifty meters is said to be the minimum requirement for an ecological overpass to attract local wildlife, the Mandai Wildlife Bridge, was initially planned to be only thirty meters wide. After environmental groups raised concerns along with the potential edge effect highlighted by the environmental impact assessment, it was later revised to forty-four meters (NSS).[25]

While the effectiveness of the Eco-Link@BKE is yet to be evaluated, the technological fix it exemplifies is already being celebrated and adopted as a solution for this new development project. As environmental concerns attract increasing attention in Singapore and globally, the eco-link becomes a means to manage public opinion by seemingly incorporating environmental concerns into a project that will disrupt existing ecological conditions. It suggests that developers are actively implementing measures to safeguard wildlife in the vicinity. In Franklin's words, this type of technique is deployed as a very practical way of "'containing' public anxiety" by "'building in' ethical concerns" to a development program (2001, 342 and 346). Positioned as a sensitive gesture toward the environment, the wildlife overpass morphs into an ethical technology, one that enables us to "bypass" difficult questions about our lifestyle and sustain continuous development and uninterrupted mobilities at others' expense.

While the eco-link and similar projects are the latest practices in conservation that may, to a certain extent, help to restore ecological connection, these green infrastructures are ultimately the consequence of some of the broken and disrupted movements caused by a desire for human mobility and drive for velocity.[26] In this light, they are living reminders of the detrimental effect that some human-centric ways of movement have imposed on the nonhuman world. Although wildlife crossing is mostly retrofitted, more recent development is seeing the concept provisioned for new road projects (Greenfield 2021). Here the dynamics of the anticipatory animal-vehicle collision, the insistence of road development, and the incorporation of a wildlife bridge as a mode of preparedness (to some extent) reveals the shifting and ambiguous space of eco-futuring.[27]

Anthony Clevenger, a wildlife biologist who has long been involved in various types of ecological passages in Banff, Canada, and colleagues describe the process whereby some animal users become comfortable with these paths as being a lengthy one (Clevenger, Ford, and Sawaya 2009). That is assuming they manage to find them at all. Yet, once such a path is established it can become intergenerational knowledge among the targeted users (Vartan 2019; see Clevenger, Ford, and Sawaya 2009). Indeed, urban infrastructures such as the eco-link or railway must be approached more than simply technology of spatial connectivity. As Cymene Howe and colleagues put it: infrastructure always has some "presumptions regarding the future built into it" (2016, 554). In the context of urban transportation, depending on how they are planned, with whom in mind, they may offer or close off temporal possibilities that sustain more-than-human relations.

Returning to the debate on the alignment of the CRL, the additional few minutes of travel time arising from the skirting option is imagined as an unbearable disturbance to human users, despite the fact that this route could go around the Central Reserve and thereby prevent further environmental impact on an already fragmented forest. Here, the urban transportation project is narrowly positioned as a technology that compresses space and aids development. At the same time, as discussed, the construction of the railway is in part motivated by the hope that it will calm political unrest arising from discontent over the uneven distribution of speed, one that favors private motorized traffic, among various sectors of society. Holding these two cases together, the seemingly conflicting acts of connecting and disconnecting enacted in the same reserve demonstrate the ambivalence surrounding urban mobility and the deployment of ethical technologies, prompting us to rethink infrastructure as layers of shadowy places and raise the serious need to craft new possibilities for urban movement.

According to Kroll, the lack of widespread protests over "the increasing loss of wildlife as a result of the modern car, the improved road and the lust for speed" is because roadkill is not perceived as "a deliberate or malicious act" (2015, 10). Kate Rigby and Owain Jones press further, suggesting that the indifference to roadkill bespeaks "a disavowal of cross-species kinship, communion and care" (2022, 115–116). Indeed, reimagining urban movement has important ethical implications. It not only calls for attention to the fact that animal movements (stillness) have been profoundly shaped by human actions in which life and death are at stake but also invites us to consider how our own movement (stillness) might be shaped by desires beyond sheer velocity, including that of learning to be moved and/or slowed down by animal residents.

Thus, to reimagine urban movement is to refuse a velocity-charged and human-centered approach to urban movement. This may involve troubling some of the taken-for-granted ways of approaching urban mobilities, such as challenging the premise that mass transit that privileges human comfort automatically equals sustainable movement, or the prevailing "certainty of knowing" in mitigating the environmental risks of transport projects. This involves respecting and acting with unknowability, putting "knowledge 'at risk'" and attending to "urban wildlife in ways that coproduce new assemblages of knowledge, people and wildlife" (McFarlane 2011, 221; see Hinchliffe et al. 2005).[28] In this context, rethinking urban movement requires not only developing a different approach but evoking a conceptual change in meditating the way we move that needs to sustain a multiplicity of relations and multispecies mobility.

In Singapore's pursuit of its imagined livable and sustainable city, the case of the CRL shows that mobility is often thought of as a key tool to maintain continuous economic development; and sustainable mobility is narrowly imagined as the practice of providing rapid mass transit with great comfort and lower carbon emissions, while any resulting environmental issues can be managed or ignored through the politics of invisibility that assigns particular entities to the shadows. On the other hand, the construction of the eco-link enabled by the city's growing awareness of the intersections between human and animal movements reflects a more experimental, negotiated, and situated practice of care (see Alam and Houston 2020) and a genuine, if only partial, desire to explore what others may need and to encourage a multiplicity of movement. Yet, as we have seen, this type of green path may also be subject to misuse as a convenient technological and ethical bypass.

As the country envisions and markets itself as a place to "live, work and play" for local residents, foreign talent, and tourists, other places and beings are denied in its making and placed out of sight. Situated at the nexus of a nature reserve, an underground railway, and an eco-bridge, this chapter reveals the complexity of urban transportation, in which the dynamics of sustainability are enlisted in both development- and conservation-oriented projects. In the process, the politics of visibility are engaged in foregrounding or casting into shadow some human or other-than-human movements, and the thirst for velocity in nation-building takes center stage. Moving through these divergent yet entangled issues also troubles the proposition of "livable cities," envisioned as a human-centered, efficient, and comfortable yet reductive mode of life sustained by continuous acceleration and through suppressing certain mobilities. Automobility and transport infrastructure are complex and contested processes that cannot be approached from the sole aspect of mechanistic engineering and scientific measures. Rather with some critical attention to mobility, these transportation projects, and cities at large, might be reimagined as an apparatus of care and multispecies flourishing.

By 2050, it is estimated that twenty-five million kilometers of new road will be built around the world (Laurance et al. 2014). As Singapore and many other cities outline the need to further expand transportation infrastructure, and as a certain version of sustainable mobility underpinned by a set of problematic shadowy issues is increasingly evoked, it is important to question the purpose of this constant "improvement" and continuous expansion. If the desire to move is carried out without any serious engagement with unknown parts of the city and their inhabitants, and if velocity continues to be the dominant force in directing urban narratives, a complex, multilayered way of living and interacting is at risk of being sacrificed for a perpetual flow between locations at record speeds.

Ultimately, this chapter is a call to pay attention to the various mobilities that are required to sustain diverse entities, human and other. In order for each to craft livable lives, we need to attend to the narratives of urban mobility that enable and render obvious *their interplays* or background and make invisible some movements, their priorities and needs, over others. Reimagining urban movement in a more inclusive way may offer a pathway to instigating deeper and wider discussions that take seriously entangled multispecies movements in an increasingly urbanized environment. It invites us to re-examine the current transport model, which is predominantly directed by projected economic and population growth and involves greater experimentation with urban planning and development. In doing so, urban mobility may emerge as a more "difficult" and less straightforward practice, yet at the same time one more supportive of permeable, ethical, and inclusive ways of moving: whose movements we intend to include, who we move with, how and to what extent we slow down, and what connections are enabled or disabled. These are ultimately questions of what type of urban future we are imagining.

4

WHAT COMES AFTER WATER?

During my time in Singapore in mid-2018, the atmosphere was tense. The newly elected Malaysian prime minister threatened to reassess the bilateral water agreement with Singapore, renewing a decades-long water feud between the two countries. Each morning, local newspapers featured headlines outlining new developments in the renewed water feud coupled with in-depth analysis of the issue: Will Malaysia cut off the water supply to Singapore? What is the current water provision strategy in Singapore? Water war brews. News channels and social media echoed each other, spewing out inflammatory comments from each government as well as any signs of negotiations. Most importantly, the media blitz centered on Singapore's capacity for domestic water production to assure the public that the city-state already had the ability to achieve self-sufficiency in water supply. However, self-sufficiency is not at present "necessary" due to the cost. In addition to seawater desalination, Singapore's own brand of high-grade reclaimed water using advanced membrane-based technologies has been widely celebrated as the way forward in domestic water supply. This supply is aptly named NEWater.

Urban researcher Maria Kaika's work on water, primarily in the context of the West, refers to the construction of big dams and water infrastructure in the twentieth century as "the apogee of modernity's Promethean project" to tame nature (2005, 141).[1] Visiting dams, Kaika says, "was a very

4.1   NEWater Visitor Centre. Reverse osmosis is a key step to purify the recycled water. *Source*: photo by author.

popular activity in the early 20th century until the late 1960s ... with people traveling hundreds of miles away to bow in front of dams in the middle of nowhere" (39). Later, worshipping mammoth engineering marvels was gradually replaced by a nonchalant attitude to naturalized domestic water. In Singapore, NEWater seems to be reinvigorating this sense of spectacle.

The NEWater Visitor Center conducts daily tours, inviting visitors to admire the production process of NEWater and promoting water sustainability (see figure 4.1). My visit to the site, just a few months before the revived water feud between the two countries, coincided with a visit by a school group from Thailand who were learning about water conservation. Over the course of the orchestrated tour, we were in awe of the stringent and layered stages of water purification techniques including microfiltration, reverse osmosis, and ultraviolet disinfection. We also grew concerned about the intense water scarcity faced by Singapore and the rest of the world, and finally we were relieved by the prospect of a sustainable water future made possible by technology.

To my surprise, although we became familiarized with advanced technology and saw various parts of the machines that produced this lifeforce, the visible presence of actual water itself was minimal throughout the visit. Only at the conclusion of the tour, each visitor was given a bottle of 100 percent NEWater to take home. I recalled Stacy Alaimo's comment that climate change initiatives by various global organizations including the United Nations "are strangely devoid of nonhuman creatures" (2016, 176). Similarly, in this high-tech water center, water is strangely missing, or to say the least, backgrounded by the work of human ingenuity in sustaining a water supply. What becomes clear is that the key narrative framing frequently surrounding both techniques—dams and NEWater, whether they function at macro or micro scales—is centered on the "power" of humanity to tame water.

With almost no natural freshwater sources, water is a national security issue in Singapore. Currently, the country imports around half of its water from neighboring Johor, a state in Malaysia, under a 1962 bilateral agreement. Since the city-state's independence from Malaysia in 1965, water has undergone a series of transformative roles. From being instrumentalized to shape Singapore's nation building, water became a bargaining chip for the Malaysian government in political negotiation and now has a new incarnation as NEWater. In recent years, the weather patterns in Singapore and Malaysia have become more volatile due to climate change. At the same time, the water quality of the Johor River in Malaysia (not a transboundary river) continues to deteriorate as a result of overdevelopment, population growth, agriculture land use, and inadequate water management.[2] With the current water agreement expiring in 2061, and more frequent extreme weather conditions including extended droughts, Singapore is determined to achieve self-sufficiency in water supply. In 2011, the Public Utilities Board (PUB), the Singaporean statutory board under the Ministry of the Environment and Water Resources, announced its new water strategy that involved a dramatic shift from importing water to domestic water production.[3] It is projected that by 2030 Singapore's domestic water production including seawater desalination and NEWater will meet 80 percent of the country's water usage. The remainder will be provided by rainfall collected in domestic water catchment areas and imported water. The goal is to achieve full water autonomy by 2060.

In 2014, the Southeast Asia Region endured a record-breaking dry spell. Yet, even as Malaysia imposed water rationing amid disruptions to its farming and fisheries, in Singapore the government assured the public that no rationing would be required thanks to sufficient domestic water provisioning through desalination and NEWater (Hansard 2014).[4] In this context, Singapore's insistence on water independence is more than a mere tactic to show political confidence before Malaysia, but also reflects a growing trust in and reliance on water technology and humanmade water as a superior source for stability and assurance. This approach is also aligned with the country's larger eco-modernization framework. To date, the small city-state already has five NEWater plants and five desalination plants.[5]

With a well-recognized integrated water resource management and an upgraded ability to produce water, Singapore has positioned itself as a global hydro-hub in water technology and policy, and, as it has sought to do in many areas, a leader in sustainable water solutions. Singapore's water story—turning its vulnerability to strength—has been widely acclaimed and researched.[6] Amid the escalating effects of climate change and increasing global water crises, this chapter pursues another line of inquiry that focuses on the country's intensification of technology-driven water and its concomitant social, environmental, and cultural impacts. What version of the future is being opened or closed for whom if seawater desalination or NEWater is imagined as the dominant mode of water provision, advocated by the state as the future-proof way of achieving water security, a "climate-independent" water source sheltered from geopolitical uncertainty and untethered from meteorology? Thinking with and responding to these concerns, this chapter puts forward the concept of *decoupled water*, describing and capturing a water imaginary that seeks to separate, extract, and produce water in an eco-modernized sense. I take the terminology of "decoupling" from this literature. To eco-modernists, radical decoupling of humans from nature enabled by technology is the key to human well-being, continuous economic development, and environmental protection (Asafu-Adjaye et al. 2015). Drawing on Plumwood's (1993) concept of hyper-separation, this chapter explores how decoupled water is an *illusory* yet dangerous project that seeks to decouple water as a part of nature that is positioned as increasingly uncertain, or uncontrollable due to environmental degradation.

This chapter is structured in four parts. The first section examines the current water crisis and Singapore's energy production, an inextricable aspect of water. The next section explores Singapore's heavily modified waterscape. How have the rivers and ocean been backgrounded, seen as mere resources, empty spaces, or threats to be secured by seawalls, and in turn impact human-water relations? I examine how deliberate segregation may induce an imaginary of decoupled water and its deep social and environmental implications. In the third section, I propose that Singapore's yearning for water autonomy and its technological pursuit of an ever-available flow of water might be a "cruel optimism" that is, as described by cultural theorist Lauren Berlant, "the condition of maintaining an attachment to a significantly problematic object" (2011, 24). Here, I point to the spatial and temporal dimensions of this cruelty to broaden the discussion. The final section of the chapter turns to some alternative ways of relating to water, from mangroves' interstitial worlding to the Active, Beautiful, Clean Waters program. The practices that seek to take the agency of water and their relations with other elements more seriously offer some hope in circumventing hyper-separation and intervening in the cruel optimism of a precarious modernist notion of sustainable water provision.

THE WATER CRISIS

Water shortage, of course, is not limited to Singapore. In *Global Risk 2015*, published by the World Economic Forum (2015), freshwater scarcity is ranked as the top risk in terms of impact over the coming decade. As uncertainty over fresh water grows, it seems that even tamed modern water is no longer sufficiently secure as a water source. There is a need for more guaranteed and controlled water. As the World Health Organization (WHO) states, "Desalination is increasingly being used to provide drinking-water under conditions of freshwater scarcity" as "competing needs for water intensify along with population growth, urbanization, climate change impacts and increases in household and industrial uses" (2011, 1).

In the past decade, the desalination industry and its processing capability have grown substantially. By 2020, there were approximately seventeen thousand desalination plants around the world, growing at the rate of 7 percent annually since 2010 (Eke et al. 2020). This is occurring despite the fact that this technique is highly energy intensive and financially costly. Unsurprisingly, desalination is mostly popular among wealthy countries or regions.[7] It is seen as the "most valuable ally for obtaining water where there is scarcity" and "unlimited" as it guarantees supply regardless of meteorological conditions (Swyngedouw 2013, 263; 2015).

According to water historian Jamie Linton, water has morphed from a mythological element to a homogenous resource in the modern era. In Linton's account, "modern water" focuses on the understanding and representation of all water "as an abstract, isomorphic, measurable quantity that may be reduced to its fundamental unit—a molecule of H_2O" that is outside of any socioecological relations (2010, 14). Positioned as a conveniently abstract resource, water is fully commoditized and intensely hierarchized through pricing (Kaika 2004; Swyngedouw 2007; Gandy 2014).[8] The recurrent tension of the water saga between Malaysia and Singapore is largely attributed to the pricing scheme that is deemed to be unfair by the Malaysian government. Under the 1962 Water Agreement, the raw water Malaysia exports to Singapore is priced at RM0.03 for every one thousand gallons, while the treated water that is then imported back from the city-state is priced at RM0.50. In Singapore's account, the price difference is fully justified; in fact, at RM0.50, it covers only a fraction of the costly and complicated water treatment process (Singapore Government 2018). Decades of conflict between the two countries over water is an example of the territorializing of water, backgrounded as mere resource or foregrounded in political maneuvers or economic conflicts.[9]

As I have examined in previous chapters in the development of Singapore, the ideology of "vulnerability" and "survival" has been consistently and carefully evoked to act as a motivating force for the country to compete and succeed. In the context of water, building on an already deep-rooted sense of resource and space limitation, scarcity and uncertainty are further emerging as key issues in the face of climate change. This includes a shortage of clean water or imminent risk of sea-level rise. These issues also inject new energy into the aforementioned siege mentality that allows the government to embark on further development in the name of national security.

However, Andrew Biro, drawing on environmental historian Donald Worster, argues that "the experience of scarcity" is "strongly culturally mediated, if not in fact culturally produced" and in turn "drives the quest to achieve a greater mastery over the natural world" (2013, 169).[10]

In Singapore, there is a stark disjunction between the narrative of scarcity and the actual experience of water shortage. The state's hubris as a water-maker seems to make water provision less an issue that affects daily life in Singapore than a policy instrument to pass or accelerate certain acts.[11] As generational Singaporean leaders persistently reiterated, "All other policies would 'bend at the knees' for our water survival" (Lee, quoted in Lee [2013]). Meanwhile the city-island-state shows no hesitation in pursuing further economic growth, and no intermission in the progress of land reclamation from the sea. Likewise, it presses ahead with water technology development to meet its growing need for fresh water. As Karen Bakker explains, modern water has become "literally, a lubricant for industrialization, urbanization, and agricultural intensification" (2012, 618). Interestingly, although NEWater has been tested as being well within WHO standards for drinking water and accepted as such by Singaporeans, only one percent of it is actually blended into the reservoir as potable water.[12] According to the PUB, as the purity of NEWater is higher than tap water, it is mainly supplied to water-intensive industries such as "power generation and petrochemical industries" and "commercial and public buildings for air-con cooling towers" (Lee and Tan 2016, 612). In other parts of the world, desalinated water is produced for drinking water but also heavily used for industrial purposes, cooling electrical generation units, and tourism (Meerganz von Medeazza 2005).

With reference to the port of Rotterdam, Thom van Dooren suggests that the port can be seen as "an 'engine' for the global patterns of production, trade and consumption" and a key driver of the Anthropocene (2019, 21). As I walk along the completely transformed coastline, the view of interminable rows of container ships in the middle of the sea breaks up the vista of the ocean and indicates the Port of Singapore as a more powerful "engine" of the era of the human. In 2013, "27% of the world's maritime oil trade and 17% of the world's total oil production sailed through the Straits of Malacca," Singapore's main shipping channel (Barr 2019, 75). Today, the city-island has the third-largest complex of petrochemical refineries in the world. Jurong island, one of the southern islands just off the coast of Singapore, is now home to hundreds of petroleum, petrochemical, specialty chemical, and supporting companies including ExxonMobil and Shell. Each day, the thirsty industries on this humanmade island devour forty million gallons of water, about one-tenth of Singapore's total water demand.[13] It is these industries that contribute to the country's strong financial position, affording it the luxury of expensive water production along with its associated CO_2 emissions.

Although water has always been positioned as an existential crisis for the country, Singapore also almost entirely relies on imported energy (although it seeks to mitigate the risk by diversifying its energy sources). In the context of the North American waterscape, Greta Gaard (2001) discussed the damage of damming for electricity that has caused long-term issues. In this case, water has been exploited to produce energy. In more recent years, along with the promise of technological water is the surging demand for energy. Indeed, both desalination and NEWater require intense energy consumption to keep the tap flowing. The continuous growth of Singapore's total electricity consumption in past years is due in part to the intensification of its water production (Vincent et al. 2014). Such an increase inevitably makes the country vulnerable to the energy market. In this light, attachment to the vision of water-autonomy grounded in an expensive and energy-intensive mode of water production is not, as it is often framed, the only way forward. Rather, the state has chosen this path over that of the other constraints for now. The rise and fall of Hyflux, the company that owns the largest desalination plant in Singapore and Southeast Asia, exemplifies the interconnection between water and energy. As water is endowed with the aura of a golden commodity, Hyflux's operating model of dual desalination plant and gas turbine power plant was once hailed as the future of water production (Whitington 2016). In early 2018, Hyflux abruptly applied for court protection to begin a reorganization of its debt, leaving the public in shock. The company attributed its financial downfall to the prolonged low wholesale electricity price.[14]

Berlant suggests that the failure of infrastructure, or the systems of reproduction, reveals "the glitch of the present as a revelation of what *had been* the lived ordinary. . . . When things stop converging they also threaten the conditions and the sense of belonging" (2016, 403, italics in original; see also Star 1999; Larkin 2013). The demise of Hyflux provides a glimpse of this glitchy system, demonstrating that the seemingly autonomous water production and the promise of water technology is inextricably situated in relation with many other issues such as volatility to the energy market and the uncertainty of geopolitics, as well as the water infrastructure. It becomes a powerful reminder that infrastructure "not only constitutes the form and performance of the liberal (and neoliberal) city but also frequently punctures its performances" (Anand 2017, 6). Moreover, it reveals the precarity and impossibility of the rhetoric of self-sufficiency and invulnerability for nations and companies to which Singapore is nevertheless attached, and thereby gives rise to a cruel optimism that I will return to in a later section.

Despite the inextricable link between its oil refinery industry and water production and the fact that energy and water consumption are on the rise, Singapore is eager to establish an image of itself as a cleaner energy user. The government proudly states that "95% of Singapore's electricity is generated using natural gas, the cleanest form of fossil fuel" and will continue to do so in the near future while looking for other resources (Energy Market Authority 2020). Urban researcher Peter Newman (2019) suggests that the current energy consumption in Singapore is not sustainable as the city-state lacks a strategy or investment in renewable energy while merely focusing on improving energy efficiency.[15] In addition, not only is natural gas a fossil fuel but it carries its own set of issues. These include high fugitive methane emissions as a result of "losses, leaks and other releases of methane to the atmosphere that are associated with industries producing natural gas" (CSIRO 2017). Singapore's lack of interest in controlling energy or water consumption, in particular in its industrial sector, also remains a key issue.

RIVER, OCEAN, AND DECOUPLED WATER

During my time in Singapore, I wandered the city and talked to the locals, hoping to collect river stories. In researcher Stephen Dobbs's words, the Singapore River "was much more than a commercial artery; it was a place where real people lived, worked, and died" (1994, 269). Yet, with intense urbanization, there seem to be few "live" memories remaining of the Singapore River. Officially memorialized as a body of water dirty as a drain and reborn after a decade-long, state-led river cleansing project, the river does not seem to evoke much cultural or spiritual sentiment, as rivers sometimes do in other cities.

Occasionally, I was referred to the sculptures along the river. This art series called "People of the River" was commissioned by the Singapore

Tourism Board in an attempt to create reminiscences about certain pasts. Among the artworks, a number of locals told me that the sculpture that resonated most with them was named "First Generation" by Chong Fah Cheong (2000) and depicted a group of playful young boys in the act of jumping into the river. The sculpture captures a nostalgic past, before the river and its surroundings were completely gentrified into a tourist site, high-end entertainment areas, and government-sanctioned cultural venues (discussed in chapter 2). At the same time, this representation of a joyful moment in the past seems to conflict with the official government narrative, which tends to focus on the once dreadfully polluted state of the river and the transformative role of the great Singapore River clean-up project (Tan, Lee, and Tan 2016).

Indeed, the cleaning up of the Singapore River was hailed as giving the heavily polluted river a new life. In literary scholar Kira Alexandra Rose's account, the clean-up "marked a material and symbolic break" with Singapore's colonial past, redefining the city-state's future-oriented vision (2021, 825). During the clean-up movement, approximately 26,000 families, 610 pig farms, and 500 duck farms were removed as they were identified as the main contributing source of water contamination (Joshi, Tortajade, and Biswas 2012). Thousands of hawkers and vegetable wholesalers along the riverbank and other backyard industries were also removed. Lighters (a type of flat-bottomed barge), once a common sight on the river, were also removed to achieve a visually appealing river (Usher 2018). Later, when the Singapore River was dammed at its outlet to the sea to become part of the Marina reservoir, it flowed more slowly and no longer served as a bustling entrepot (Dobbs 2003).[16]

Since independence, Singapore has deeply invested in the expansion of water catchment and drainage. Most of its waterways have also been dammed or heavily modified into canals. To date, the water catchment areas take up two-thirds of its land surface. Very few freshwater streams survived this complete colonization of waterways. Despite their significant ecological function, the remaining streams continue to be at risk of being straightened and canalized to make way for development work.[17] In the name of protecting this essential resource from contamination and exhaustion, canals were created to separate the water from the population (Usher 2018, 327; see also Liao 2019). Not only was the route of the water

standardized, and its rhythm controlled, but water has also been desig-
nated by law as the exclusive property of the PUB (Usher 2018, 321). This
ruling includes not only river water but also rainwater. Any unauthorized
rainwater collection is illegal, as each drop of water is deemed to belong
to the state. In more recent years, people may collect rainwater for non-
potable use subject to various guidelines. As water's hydrological flows
and connections with other waters are controlled, also in the process is the
water-human connection heavily regulated or severed.

However, not all urban inhabitants follow this order. Although the
closed drainage system with concrete covers in Singapore has been effective in isolating the water source from humans, it soon became a refuge for
mosquitos. Faced with recurrent outbreaks of dengue fever due to mosquitos, the city-state was forced to consider a softer approach toward controlling water, including a mass removal of concrete drainage covers (Usher
2018). In recent years, researchers have meditated on the ways in which
mosquitos and their plasticity toward humans' ongoing eradication efforts
have shaped urban topography around the world. For example, Gandy
(2014) traces the inextricable and shifting relations between mosquitos,
malaria, and the process of modernization in Lagos, Nigeria. In Tess Lea's
work, she credits the mosquito as "a powerful urban planner" in shaping
the development of the Australian city of Darwin (2020, 84). For instance,
Lea explains that, in the case of Palmerston—a satellite town in Darwin
built for lower-income earners and welfare recipients—serious concerns
for mosquitos and related diseases have seen the suburb end up being
equipped with "the most advanced underground pipes and culverts," disrupting the continuum of "affluence and amenity" (2020, 96–98).

Among these diverse urban relations woven by humans, the behavior
of mosquitos, and ongoing development work, what I seek to highlight
is that, in O'Gorman's words, people need to learn and have "learned to
see environments differently" through a more-than-human lens (2021,
76).[18] Attending to some of the "awkward" creatures in the urban when
"togetherness is difficult" and multispecies "vulnerability is in the making"
(Ginn, Beisel, and Barua 2014; see also Ginn 2014) also challenges a seemingly alluring narrative of multispecies coexistence in envisioning urban
sustainability. These include mosquitos as well as the macaques discussed
in the introduction, whose presence are not welcome in close proximity to

humans' homes. In short, these "difficult" animals are always, and increasingly so, a part of the city.

Later, since 2004, the PUB has extensively reformed its public relations strategy to increase the social appeal of water (Tortajada, Joshi, and Biswas 2013, 123). Yet, unlike the spontaneous, intimate, and open watery past captured by the "First Generation" sculpture, the attempted transformation of human-water relations is measured and planned. As water is alienated from its aqueous connections, it is linked to human amenity in other ways. As Fiona Allon writes: "Today, the display of water (in swimming pools, gardens, water features, etc.) functions more than ever as a symbol of cultural capital and distinction and as a key indicator of wealth and social power" (2020, 238). In Singapore and many cities around the world, proximity to a body of water has become one of the attributes of prestigious housing estates, thereby morphing into other forms of commodities.

With intense urbanization, the imagining of and the distance to the sea has also been heavily modified. By now, the topology of the Singapore coastline has completely changed as a result of one of the most globally aggressive land reclamations.[19] While walking beyond Beach Road in Singapore, it would sometimes occur to me that the vast area I had just strolled through, including some of the most highly valued developments, was all reclaimed land. In writer Samanth Subramanian's (2017) words, "Beach Road, in the island's belly, at one time had a self-evident name; now it reads like a wry joke, given how much new land separates it from the ocean." The land reclamation for urban development and the damming of all the major estuaries for freshwater reservoirs also wiped out 90 percent of the mangrove forests that once connected the salt and fresh water (Lai et al. 2015). Historian Miles Powell (2019) suggests that the extensive land reclamation and resultant displacement of local communities has had a significant ecological and social cost, and is responsible for Singapore's, albeit an island nation, profound cultural disconnection from the sea. Indeed, the impact of the erasure of the environment is never singular, and forgetting is always natureculture (chapter 2). In the face of the increasing threat of sea-level rise, over 70 percent of Singapore's coastline has been enclosed by seawalls or rock slopes, for protection. As a low-lying country with over half of its precious freshwater reservoirs situated near the sea, the fear and risk of rising sea levels and seawater intrusion into reservoirs is real. Yet, the extent to which the country has augmented and encroached on the seaspace through land reclamation, conversion of estuaries into water supply infrastructure, and damming has removed the prior conditions of coexistence and continuity.

In the face of climate change, eco-modernists maintain that there is a need to consciously accelerate and intensify the decoupling process. This involves intensifying human activities such as urbanization, farming, energy extraction, and desalination, yet restricting these activities to as small a footprint as possible. According to this logic, such a process might allow people to mitigate climate change and alleviate global poverty, while leaving as many (relatively) undamaged wild places as possible (Asafu-Adjaye et al. 2015; see also Blomqvist, Nordhaus, and Shellenberger 2015). As the river and various types of water are modernized and contained in canals, and as the sea is pushed away due to its perceived risks to some humans' flourishing, the techno-waterscape grounded in fragmentation, isolation, and separation gives rise to *decoupled water*. While this process of decoupling is generally presented in a positive light by eco-modernists, my understanding of this process is tightly tied to the concept of hyper-separation. In Plumwood's (1993) critique on human-nature dualism, she terms the ultimate separation from nature as hyper-separation, a form of radical exclusion and domination. But, for Plumwood, this kind of separation is always also illusory in important ways. As she points out, "the resulting delusions of being *ecologically invulnerable*, beyond animality and 'outside nature' lead to the failure to understand our ecological identities and dependencies on nature" (2009, 117, italics in original).

In the case of Singapore, decoupling, emboldened by a sense of technological supremacy, encourages the notion that any part of nature that is uncontrollable or unusable needs to be denied or secreted away. By these means, the city-state is further separated from its ecological surroundings. As Singapore shifts its position from one of being water dependent to one of a water maker, it expounds a new imaginary of water future in which it seems possible to overcome water scarcity provided there is sufficient technology, wealth, and energy. For example, NEWater, celebrated as a savior for Singapore's water supply, has been included in educational material as a crucial water source to secure the city-state's water future and prosperity (Irvine et al. 2015). One of the key dangers of this kind of decoupling work is that it "distorts our perceptions and enframings in ways that make us insensitive to limits" (Plumwood 2009, 116). Further, it makes the imagining of water autonomy (although illusory) appear real and necessary.

The imagining of water in an eco-modernized sense is also reflected in the naming of the water. On the one hand, it can be understood that the branding of NEWater softens the image of recycled water to increase public acceptance. On the other hand, the naming of NEWater and the language used here is highly suggestive. It indicates that the water is produced anew by the state and humans. Recalling my visit at the NEWater center, it was clear that the minimal presence of actual water and the maximization of the infrastructure is doing a kind of affective work that prevents visitors from experiencing water or understanding "our embeddedness in and dependency on nature" (Plumwood 2009, 116; Knox 2017; Appel, Anand, and Gupta 2018). At the same time, the construction of seawalls produces a physical separation between land and sea and a conceptual detachment from nature that enables the ocean to be seen as merely a free resource, or a dangerous terrain that needs to be held at bay.

In the midst of a new round of water crises, water is simultaneously foregrounded as a life force and deprived of its agency, in favor of a celebration of technological water offering stability and security. As the attempted process of decoupling seeks to work around ecological limits, it also downplays the socio-environmental impact of its own production, many aspects of which have been transferred to other countries. A central assumption of ecological modernization theory is that continuous economic growth paired with a techno-centric approach should no longer require an unsustainable exploitation of resources (Cohen 1999; Mol and Sonnenfeld 2000; Lidskog and Elander 2012). However, in reality, this has not translated into practice to any great extent. In Singapore and beyond, the consumption of minerals continues to rise despite increased efficiency in the use of resources (IEA 2019).[20] In sociologist Elizabeth Shove's account, "the point is that efficiency is itself part of the problem" (2018, 780). This is a salient point I will discuss further in a later section.

Although Singapore lacks natural water sources, the government states that the water industry has "created some 14,400 good jobs and economic value-add of over $2.2 billion annually. . . . Singapore is now one of the best and busiest hydrohubs, where companies come to prove their patents and scalability in a live water system" (Zulkifli 2019a). In this future, the water sector has become a new profit center of the "blue gold" (Allon 2020, 233). As Singapore exports both its water technology and the products of industries that use water, it facilitates a hydrocapitalism that not only contributes to the accumulation of capital but also takes advantage of scarcity (see Neimanis 2017).[21] In this light, decoupled water is an apparatus both for overcoming ecological limits and also satisfying the relentless pursuit of consumption.

Positioned as a modern water maker, the future that Singapore seeks to secure is supported by a hydro-techno-capitalist loop that not only insists but ultimately relies on and benefits from "business as usual." These layered attachments brought about through social, political, and economic mechanisms construct and sustain a certain mode of life that is increasingly difficult to escape. In light of the rising popularity of such technological water processes as desalination or NEWater, there are more questions, both existing and emerging, that need to be addressed and/or reexamined.

SUSTAINABLE WATER AS CRUEL OPTIMISM

Marina Barrage, a dam built across the 350-meter-wide Marina Channel, is Singapore's fifteenth water reservoir. Near the edge of the barrage, a long bridge erected on nine steel crest gates divides the once continuous waterscape. The official sign on the barrage says that these metal giants keep out the seawater. On the left is the untameable, bitter, and undrinkable salt water, on the right is the calm and protected reservoir, a popular place among the locals for water sports.

The rooftop of the Marina Barrage is an unexpected public green space offering a panoramic view of the Singapore skyline, overlooking the opulent Marina Bay Sands Hotel with its spectacular surfboard-shaped deck, the futuristic Gardens by the Bay, and the South China Sea. It is also a popular space among locals for picnicking or kite flying (see figure 4.2). Not too far from the Marina Barrage, Singapore's fourth desalination plant started operating in July 2020. In the account of the PUB (2023): "[U]naffected by tides, water level in Marina Reservoir is kept constant all year round. This is ideal for all kinds of recreational activities such as boating, kayaking and dragonboating." This scene seems to have encapsulated what

4.2 The rooftop of the Marina Barrage. People are gathering for the air show rehearsal for the National Day parade. The green roof is dubbed the best viewing platform in the city. *Source*: photo by author.

Singapore envisions in regard to water: a fortified, ecologically invulnerable space with humans at its center.

As I sat on the roof of the barrage, my mind also wandered to some of the many politically and emotionally charged water scenes around the world, from the Israeli–Palestinian water conflict to the battle for Indigenous water rights in North America, and to the devastating situation in the Murray-Darling Basin in Australia. I thought about how access to reliable, clean water can be an extreme privilege in cities such as Cape Town (Green 2018) and Mumbai (Graham, Desai, and McFarlane 2013) that have suffered drought. In contrast to the dire situations in those parts of the world, the water scene in Singapore appears remarkable. Here, akin to its calmed reservoir protected by the barrage, the tension of water issues is tempered by technoscience. Sustainable water provision is discussed in international conferences and through transnational public–private partnerships.

Cultural theorist Astrida Neimanis (2017) has argued powerfully and carefully about water and its movement through multiple water cycles, which always also include human bodies, materially and imaginatively. Indeed, I am fully aware of my "watery embodiment" and the continuity of the sea and the land. Yet, seeing the barrage with its steel structure that securely separates the community from "dangerous" nature while providing a calm reservoir and an expansive green space for leisure, I momentarily feel my distance from water. From the seawall to the barrage, there are clearly some effective, affective, and in turn problematic attempts to overcome continuity and conjure a particular narrowed sense of security. How might these heavily invested and designed imaginaries cast their powerful force onto the public? Watching the sun slowly set over the solid gates of the barrage, I felt increasingly unsettled by this seemingly harmonious scene underpinned by a hyper-separated human-water imaginary. Could this be the way we live in the future in which (some) humans take ever-increasing control of the hydro cycle, recycling all the water to make it so clean that only machines are deemed fit to devour it? Who could afford to enact this good life, and at what and whose expense?

Berlant describes the phenomenon of cruel optimism, defined as the relation that "exists when something you desire is actually an obstacle to your flourishing" (2011, 1). Although Berlant explores cruel attachment in the context of unachievable fantasies of "the good life" such as the promise of upward mobility, enduring intimacy, and political security, and does not explicitly engage with environmental issues, her concept provides a provoking and useful framework for thinking about the paradoxical issues of water scarcity, environmental damage, and the desire for an ecologically sustainable and invulnerable future.

In Singapore, an imagined secure and hydro-capitalist sustainable life enabled by the attempt to decouple water has become the cruel object/ object of desire that the state clings onto. Indeed, around the world, "whereas traditional hydraulic engineers lamented the fading prestige of their terrestrial solutions," the new generation of eco-modernists embraced the prospect offered by desalination (Swyngedouw 2013, 265). As sociologist Richard York and Eugene A. Rosa put it, late modernity, to ecological modernizers, "offers promise that industrialization, technological development, economic growth, and capitalism are not only potentially compatible with ecological sustainability but also may be key drivers of environmental reform" (2003, 274). The optimism, Berlant explains "becomes cruel" as "a person or a world finds itself bound to a situation of profound threat that is, at the same time, *profoundly confirming*" (2011, 2, italics my own empha-sis). And so, how might the attachment to decoupled water actively inhibit the goal of achieving a sustainable environmental life?

First, some of the decoupling and intensification processes, such as water desalination, have significant environmental impacts. Researchers in a range of disciplines are increasingly concerned with the effects of the waste generated by large-scale desalination, in particular the ecological impact of the hyper-saline brine—the afterlife of desalination—discharged into the sea.[22] A recent study on the global outlook on desalination and brine highlights that there is a significantly higher amount of brine production than was previously quantified (Jones et al. 2019). In 2017, facing a pub-lic inquiry on the potential ecological impact of desalination, the answer from the spokesperson of PUB was vague, "Because the ocean is so large, and with replenishment of fresh water from rainfall, the salinity impact of Singapore's desalination plants is negligible on a global scale."[23] This type of thinking embodies what Alaimo has referred to as a "persistent (and convenient) conception of the ocean as so vast and powerful that anything dumped into it will be dispersed into oblivion" (2012, 477). As reported by the local media (Tan 2017), the PUB spokesperson further adds that the desalination process is being carefully studied, with mitigation planned and implemented to ensure the localized impact is within tolerable limits.

In a study on the Sydney desalination plant in Australia, researchers con-clude that there is no major impact on marine communities from brine as a result of the good engineering and in turn the efficacy of high-pressure dif-fusers in ameliorating hypersalinity (Clark et al. 2018). The study also sug-gests that any ecological impacts "appeared to be driven by the increased flow created by the high-pressure diffusers" (Clark et al. 2018, 767). How-ever, they clarified that "These ecological changes may be less concerning than those caused by hypersalinity, as the currents were still within the range that marine communities experience naturally" (Clark and Johnston 2018). Although Graeme Clark and colleagues briefly mention the need to better understand these impacts, their findings have been seen as very encouraging and reported as such in the context of the growth in desalina-tion in a time of increasing drought and climate uncertainty (Snell 2018). At the same time, others are less optimistic. After analyzing various possi-ble brine disposal methods, Argyris Panagopoulos and Katherine-Joanne

Haralambous (2020) highlight that brine treatment remains a central issue in terms of the adverse environmental impacts of desalination. They further suggest that the energy consumption associated with the technology for brine treatment needs to be taken into consideration. Researchers are also concerned about the growing intensity of the environmental impacts as a result of the expanding application of desalination around the world (Elsaid et al. 2020). As I discussed in chapter 3, there is a profound belief in some of the human-centric urban planning practices that environmental impacts or risk can always be "mitigated" through science and engineering.[24] Reflecting on these contested views, much more work is required to understand the immediate and latent ecological implications.[25] While we are still learning these effects, one thing that seems to be more certain is the potentially more radical increase in the reliance on these types of technological water.

In addition, like many other techno-centric eco-modernization solutions—such as the fuel cell (Hultman 2013)—the production of desalinated water and NEWater are energy and capital intensive, contributing to significant CO_2 emissions, not to mention hydration of other industries including petrochemical facilities. Despite the fact that the small island already uses 430 million gallons of water every day, its projected water usage by 2060 is set to double, of which 70 percent is from the nondomestic sector. If petrochemical industries, wafer fabrication plants, and commercial buildings are seen as the "engine" of the country that enables the production of capital, NEWater that is used for industrial and air-con cooling purposes is literally cooling and sustaining this otherwise overheated engine.[26] Thus, intensification continues to satisfy the voracious appetite for clean water and further exacerbates some of the environmental issues it seeks to resolve.

Schneider-Mayerson suggests that, in the face of climate change, Singapore's extreme wealth positions it in complete contrast to some other nations: "As a wealthy nation with a tradition of environmental engineering, a strong centralized government, and the technological capacity to adapt," Singapore "might provide a glimpse of the future for those lucky enough to survive the rising tides" (2017, 166). It seems the insistence of clinging onto a precarious good life narrative that is harmful to the environment is in part imbricated with those on whom this harm is inflicted.

Compared to Malaysia, which has long weaponized water to exert domination over another country, on the surface Singapore's technological response to water scarcity seems to carry minimal impact on others.[27] Yet, the environmental impact of water production and the industries that generate the capital to sustain the costly water production are often unevenly distributed, with some countries carrying a greater burden of the environmental damage than others.

In examining Singapore's water usage, scholar Davy Vanham (2011) draws on the concept of "virtual water" to identify the "hidden" flows of water embedded in the production of commodities (see also Allan 1998). Vanham highlights how, as Singapore imports water via food and industrial products, it relies indirectly on overseas water sources. This net import equals around "20 times the amount of its current total blue water supply" (226). Viewed in this way, the island city-state, with its global hinterland, can implicitly transfer much of its water usage to other countries. The politics of invisibility reveals how the development of cruel optimism connects the spatial and the geographical. Given that agriculture is one of the largest users of water, the concept of virtual water becomes more poignant when Singapore's importation of food is considered. Many imported foods in Singapore, including its staple, rice, are from countries such as Vietnam, Malaysia, or Thailand that are already experiencing various water issues. As one of the few countries sufficiently wealthy to engineer solutions to water shortages at least on a temporary basis, the seemingly post-political technomanagerial process of producing decoupled water is always already a political work. It transforms one's own relation with water as well as that of others. The resultant environmental injustice is a choice that the country has made and in turn imposed on others, either wittingly or unintentionally.

In addition to the environmental impacts of water production, the second key aspect of this process of cruel optimism centers on the way in which water governance tends to be heavily biased toward curbing *domestic* water usage, while sidelining larger, systemic, issues. Significantly, while Singapore's nondomestic sector currently accounts for 55 percent of the country's total water usage and is expected to increase to 60 percent by 2030 and 70 percent by 2060 (PUB 2018b), its demand-management policies place a much heavier emphasis on the domestic water sphere. Along with ongoing water conservation campaigns targeting domestic water users, the state launched a range of promotions on water efficient appliances and recently a smartphone application that monitors household water use. Although some water-saving measures target nondomestic usage, the government ultimately supports ongoing growth of usage in industrial sectors as part of the growth of the economy.

Like Singapore's climate change policy that focuses on energy efficiency and green growth (MEWR and MND 2014), the primary focus on the water issue is to advance water technology. The aim is to reduce the energy required for water production and thereby meet the increasing water usage required to continue mass production (Vincent et al. 2014). Some scholars point out that the approaches to water planning in Singapore and other developed countries reflect a tendency toward the supply-side efforts and reliance on technology (Ong et al. 2019, 134). Others have warned against the singular aim of improved efficiency, as such a one-dimensional approach may trigger the Jevons paradox, meaning that the increased technological efficiency of water production ultimately leads to greater consumption as a result of increased affordability (York and McGee 2016; Herring 2006).[28] Shove writes candidly, "The problem with efficiency policies is that they are much *too* effective, not in reducing demand but in reproducing and stabilizing essentially unsustainable concepts of service" (2018, 785, italics in original). In the case of water production, the reduction in desalination energy requirement (especially around membrane-based technologies) as a result of heavy research and development is a determining factor behind the global proliferation of desalination plants (March 2015, 236).

Moreover, Shove's thinking shines a light on the temporal aspect of cruel optimism, given that, in relation to hyper-separation, the processes of cruel optimism may inflict their suffering on various parties at different times. Indeed, a significant point of hyper-separation is that it ultimately harms both self and other (Plumwood 1993, 2009). In Singapore, in the short to medium term, the desire for a sustainable water supply based on such separations seems to be working well locally, even if the flourishing of others beyond the country's borders is already being impacted. However, this cannot last forever. This is where the cruelty, in other words, the unattainability, lies.

In Singapore's dominant water imaginary, safeguarding itself with seawalls as well as employing technological measures and capital appears to be the most reliable solution to water scarcity and climate change. By these means, the city can continue to flourish although some expensive yet convenient adjustments are required. For example, it has been necessary to revise the minimum land reclamation level from three meters to four meters above mean sea level (MEWR and MND 2016b). Of course, water scarcity is undeniably a real vulnerability for Singapore, both for the inhabitants and the economy. Yet, adhering to a sense of invulnerability through various forms of separation can lead to an illusion of freedom from ecological dependency.

There is no clear remedy or clear path to escape from cruel optimism in Berlant's discussion. However, amid the escalating water issues, sea level rise, the vicious cycle powered by hydro-capitalism, a productive dialogue between Judith Butler's (2004, 2009, 2012) more recent work relating to the dynamics of precarity and cruel optimism may open up possibilities to rethink the notion of precariousness and vulnerability. In the context of 9/11, Butler warns against the attempt to become secure by separation rather than by building more sustainable connections. Instead of trying to become invulnerable, Butler asserts the irreducible vulnerability of life and the impossibility of being autonomous: "So we have to rethink the human in light of precarity, showing that there is no human without those networks of life within which human life is but one sort of life. Otherwise, we end up breaking off the human from all of its sustaining conditions (and in that way become complicit with the process of precaritization itself)" (2012, 173).

As Kelsey Green and Franklin Ginn point out: "While Butler's work is predominantly human-oriented, her arguments about vulnerability need not be limited to the human" (2014, 153). I agree that Butler's thinking is apt in the context of a more-than-human city. She asks us to think about the particular ethos and politics underscoring "global interdependence and objects to the radically unequal distribution of precarity" based on denying a shared precariousness (Butler 2012, 170). In this case, the imagining of unlimited decoupled water through technological advancement and the pursuit of some form of autonomy is also a political project that continues to ignore (inter)relations and deny (inter)dependence. Seeing Butler's approach through a more-than-human lens may intervene in the eco-modernists' emphasis on decoupling from nature, preventing it from being put into practice solely on human terms or fully mastered.

The eco-modernization narrative, when viewed through the lens of cruel optimism, clearly adheres to the fantasy of continual growth, illusory self-sufficiency, and backgrounded nature. Ultimately, Singapore's desire to fortify itself through selective attention to the intensification of water production disregards, or at least fails to take seriously the enormous energy cost of technological water practices. It also, of its nature, involves further development and profound disconnection from wider ecological issues. Crucially, these water solutions are not merely concerned with satisfying current needs but about enabling ongoing growth of population and industry that must ultimately cause harm to all, although this harm may be experienced unevenly. There is a real urgency to intervene in this alluring and dangerously perpetuated way of thinking that belies separation from ecological damage even as it transfers it to others. The temporal and transnational aspect of the cruelty involved in adhering to a seemingly invulnerable future make it clear that the attempt to decouple is not only from aspects of nature that are uncontrolled but also the relationships and the responsibilities that come with these aspects.

HOW DOES URBAN WATER SING?

Drawing on Linton, Neimanis points out that the water crisis is indeed a social crisis as it is "largely precipitated by a social imaginary of what water is" (2017, 20). In the time of technological water, the narrative perhaps also lies in the question of where water is from, and more specifically, the way in which water is produced. The promise of ever-available water undermines the efficacy of water conservation work, even in the domestic sphere. A more recent survey conducted by the PUB and REACH (the government feedback unit) found that younger participants viewed water scarcity as "a distant concept" (Elangovan 2019). Some suggested that water security might no longer be an issue thanks to the development of water technologies such as NEWater and desalination. Here, the indifference toward water precisely points to the need to approach water governance differently.

In Singapore and elsewhere, I suggest that the work of repairing the rift between humans and their watery history, and of circumventing an attachment of cruel optimism, requires an onto-epistemological shift. It requires unlearning the fantasy of abundance promoted by capitalism: a narrative that advocates an illusion of continual growth and the ability to be self-sufficient by backgrounding nature. At the same time, there is a need to (re)learn to attend to a more relational thinking about water. How might we, as Neimanis challenges, "find ways of imagining watery nature as continuous with human nature, and therefore responsive to human actions, but in a way that does not deny waters their power and agency as collaborators" (2014, 10). Of course, the separations that have occurred in the process of urbanization and modernization have left indelible marks on the many humans and nonhumans and as a result forming ecological attachments or reconnecting is not straightforward. While this chapter has focused on urban water, in the process, I have demonstrated the need to look beyond the human-centered and progress-driven kind of demand and supply management to find potential solutions. An approach based on improved efficiency in water production or reducing domestic water consumption (although important) circumvents some deep-rooted issues that contribute to the conditions producing the cruel attachment and process of decoupling.

Along with the aspect of recognizing a more-than-human precariousness discussed in the preceding section, Stengers's work on cosmopolitics offers an expansive and inclusive approach to rethinking and reconnecting. It suggests a way to circumvent cruel optimism as it interrogates and rejects the normative business as usual kind of "'good' definition of the procedures that allow us to achieve the 'good' definition of a 'good' common world" (2005, 995) and challenge us to redefine the static knowledge that we have learned. Although decoupled water works somewhat differently, thinking with cosmopolitics offers ways to explore how water provision, water governance, and water processes might be approached differently, as well as ways to refuse the externalization of water. In accord with this understanding, my analysis points to some hybrid, more open approaches including renewed emphasis on re-naturalization of rivers and mangrove restoration. These cases might be positioned as fragile, emerging examples of other ways of relating to water. In this way, they offer hopeful alternatives into a process that may maintain other ecological relations and resituate humanity in ecological terms. They may also help us relearn the agency of water and create watery spaces for possible cohabitation.

All over the world, researchers and environmental groups are calling for a more flexible and enduring form of coastal defense than seawalls or dikes

(Cheong et al. 2013; Wong 2018). Singaporean geographer Wong Poh Poh, who has long worked with the UN Intergovernmental Panel on Climate Change (IPCC), pointed out to me in an interview that serious attention need to be paid to more diversified adaptation measures toward the issue of climate change and sea level rise in Singapore, rather than an overreliance on seawalls as a coastal defense mechanism.[29] He explained to me that rigid infrastructure such as seawalls are unable to keep up with sea level rise or react appropriately to the uncertainty of climate change. He pointed out that mangroves and other more ecologically based methods are often better adapted strategies to prevent erosion, and counter sea-level rise and coastal storms, while also restoring coastal ecosystems.[30]

As ecological coastal defense gains more attention, large-scale mangrove reforestation projects are emerging. Like the relationship between residents and the Singapore River, coastal residents related intimately to mangroves through their daily interactions (Thiagarajah et al. 2015; Friess 2017a). With only 5 percent of the original mangroves left in Singapore, ecologists suggest that the few remaining and newly planted mangrove patches offer limited ways in which people may access or engage with them, which are largely reduced to "recreation," "ecotourism," or occasional "education" sites (Thiagarajah et al. 2015). Coastal scientist Daniel A. Friess (2017b) also observed that the seedling mortality rate was high in many mangrove reforestation projects and attributed this failure to a combination of reasons including reduced photosynthetic ability in artificially replanted monocultures, poorly situated knowledge of species, and lack of engagement with the local residents (see also Lewis III 2005).

As a contrast, Friess (2017b) highlights a community-based rehabilitation project on Pulau Ubin, as a hopeful case that may inspire long-term positive ecological outcomes. Since the original mangrove forests in Ubin, an offshore island of Singapore, were cleared to develop aquaculture ponds, the island's coastline has suffered severe erosion. According to Friess, an integral aspect of the "Restore Ubin Mangroves" project, launched in 2016, has been to create a more meaningful engagement with the local island residents and the wider public. Information posters and leaflets were produced in three languages to connect with local villagers (Friess). As well as workshops on planting mangroves and the ethos of ecological restoration, the project organizes monthly walks and knowledge sharing to improve the understanding of the local environment and hydrological patterns, and to increase public participation and ongoing interaction with the mangroves (Restore Ubin Mangroves).[31] In short, the restoration project focuses on increasing the ways in which people may (re)learn to dwell in the space of mangroves and waters and be embedded more firmly in their own local environment. In this context, this approach recognizes that reconnection cannot be achieved through mechanical replanting exercises directed in a top-down approach to nature—an approach that also denies the agency of mangroves and their social, cultural and ecological meanings.

Thinking with Stengers, Houston and colleagues suggest that cosmopolitics is particularly useful for approaching the more-than-human city as it describes "constellations of diverse inhabitations"—where disparate entities: "human and nonhuman stories, relations, politics, and practices can connect to destabilize the hegemony of technocratic responses to climate change" (2016, 261).[32] Indeed, not only is it important to restore mangroves as an ecological measure for coastal defense and carbon sequestration (Friess, Richards and Phang 2016), but it is pertinent to trace and learn how mangroves, in collaboration with water, act as an agent countering hyper-separation as it connects diverse entities through openness and continuation. During my visits to various mangrove forests, I learned how these liminal beings inhabit the area where salt and fresh water meet and how they partner with water through their dense tangle of prop roots. As the tangled roots help to slow the movement of tidal waters, more sediment settles and builds up the muddy bed. Stengers writes: "'slow down' reasoning and create an opportunity to arouse a slightly different awareness of the problems and situations mobilizing us" (2005, 994). This proposition resonates movingly with the movement of mangroves and their effect. When water hesitates and "roots" in its surroundings, species socialize with each other, repairing the broken relations and forming new ones, conceptually and physically.

If technological water production, seawalls, and dams suggest a static and decoupled hydro-social relation grounded in a persistent effort to dominate watery forces, the partnership of mangrove forest, salt water, and fresh water maintains a continuity between the land and sea, transforming a rigid border into a permeable zone of entanglement. In this way it creates a much-needed bio-interstice, enacting the possibility of co-living in the area. Walking along the seawalls, it often sounds noisy as the seawater hits the resistance of the solid structure. Among the mangroves, movements and rhythms seem to be calmer as the watery forest dissipates the waves' energy while generating more opportunities for others.[33] Thinking with mangroves' way of relating to water opens up ways to (re)learn to reinhabit and resituate in a less defined yet fertile more-than-human space. Could re-encountering the agency of water, in this case, interrupt the capitalist refrain that has been directing decoupled water, and encourage a new kind of attachment in water management?

Moving from the coastal area to inland, many urban rivers are buried underground, forcefully stratified or dammed. Cecilia Chen and colleagues suggest that "the way we choose to build our cities can severely limit our understanding of water and may even encourage its forgetting" (2013, 9). For instance, the human-water relation is often channeled by or restricted to "taps and plumbing ... drinking fountains ... or waterside parks" (9). Recently, there has been a growing movement to "day light" or re-naturalize urban rivers, from the restoration of Cheonggyecheon stream in Seoul that seeks to resituate water in an urban environment, to the Netherlands' "Room for the River" program that teaches how to live with flood safely through giving the river more room (UN-Habitat 2012, 39–41; Goossen 2018). In Singapore, the PUB launched the Active, Beautiful, Clean (ABC) Waters program, a strategic initiative to re-naturalize concrete drains and canals, and incorporate rivers and reservoirs into the surrounding environment in order to create clean and vibrant waterways. The executive director of Singapore's planning center Khoo Teng Chye describes the purpose of the program in the following way: "Unlike monsoon drains, the re-naturalised canals allow children to walk up to the stream, get their feet wet, catch fishes and see all sorts of wildlife, like egrets and otters. They can enjoy the same pleasures their grandparents did in rural streams, but which their parents' generation—like most other city dwellers—rarely experienced" (2016, 60).

The launch of the ABC Waters program was accompanied by a series of public participation activities to promote an understanding of the initiative among Singaporeans. To date there has not been sufficient qualitative research to know definitively whether these projects are effective in engaging locals, and to what extent.[34] Some researchers have suggested without ongoing research on water quality remediations that the water program may become merely forms of urban parks, in part owing to the fact that the program requires expensive engineering and ongoing maintenance to establish and remain the functional and "postcard-pretty community spaces" described by the PUB (2018a, 1; see Lim and Lu 2016).

4.3 Bishan-Ang Mo Kio Park, a flagship site of the ABC Waters program. *Source:* photo by author.

It was a quiet day when I visited Bishan-Ang Mo Kio Park, a flagship site of the ABC Waters program (see figure 4.3). In stark contrast to many long and narrow concrete canals in Singapore and other cities, a lively river meandered through a wide and expansive green area. The water here clearly sings louder and more emphatically than that of the concrete drains. Compared to a dam that relies on a rigid divider, the edges here between humans, plants, and water are blurred, evoking a much more intimate feeling. The entwined water streams spread like veins, intersecting and diverging. Recalling the "First Generation" sculpture located on the bank of the Singapore River, I feel a sense of hope that perhaps there are ways for new generations of Singaporeans to reconnect with and learn from water, even if it might initially be in the form of more managed patches of water.[35] In many ways, the water program is already working in the sense that it has increased the visibility and accessibility of a less tamed form of urban water.

At a time when many sustainable water solutions are increasingly associated with intensified technological water production, these emergent examples offer an alternative aqueous way of relating and partnering that resists the externalization of water. This may in turn help us to cultivate a different set of responses and modes of relating. For instance, how might we (re)think our generational connections to tap water in different ways? To this end, these lively projects invite us to reimagine a watery relation that holds, channels, and irrigates our more-than-human body.

Importantly, I am not proposing another kind of dualism of tamed and "natural" urban water, or arguing here that regrowing mangrove forests is the ultimate solution. Such a reduction may create its own issues, such as recruiting the wetland as a form of labor in capital accumulation (Robertson 2004).[36] Similarly, experimenting with an ecological wall that is more hybrid than structural is only a partial solution for some areas. In short, nature-based solutions have to be seeded in the desire to create space and conditions for co-living, and to acknowledge humanity's rooting in its ecological surroundings. These solutions cannot simply be approached as another project to solve human issues. It is essential that they be accompanied by or form part of a systematic change. For example, it is necessary to halt, or at least to reexamine, the need that drives further land reclamations that in turn demand stronger and greater defense systems.

In 2018, the IPCC released a special report "Global Warming of 1.5 °C," in which the authors warned that rapid, far-reaching, and "unprecedented changes in all aspects of society" are required to avoid disastrous levels of global warming that would cause seawater intrusion, increased flooding, and decreased availability of fresh water. Less than a year later, in early 2019, ExxonMobil announced that it would invest several billion dollars in further expanding its oil refinery complex in Singapore. The expansion is hardly a surprise considering the history of Singapore's strong support of the industry. Singapore prime minister Lee Hsien Loong's speech at the opening of ExxonMobil's new facility in 2014 captured the country's ambivalent position in committing to environmental issues and supporting petro-capitalism: "The energy and chemicals industry is a major carbon emitter worldwide. It is the nature of the industry. . . . Singapore has committed to reduce carbon emissions . . . once the UNFCCC reaches a legally binding global agreement . . . [W]e must reduce our emissions. . . . But at the same time, I want to assure all the energy and petrochemicals companies here that the Singapore Government stands fully behind them and will continue to help them to succeed."

Without doubt, the oil giant will demand more water to hydrate its engine. In return, it will productively generate more capital to enable further intensification of water production. Singapore's strong water governance and advanced water technology has a prominent regional and growing global presence. As other countries turn to the city-state for their water expertise, what vision of the future might they swim into?

In this chapter, the concept of decoupled water is understood to be driven by the desire to separate from a part of nature deemed uncontrollable. In the process, it rejects the need to observe ecological limits and reinforces dualism. Nevertheless, the determined pursuit of a narrowed vision of water autonomy ultimately undermines the dynamics of other environmental issues as well as encouraging a deliberate separation from the resultant ecological crisis.

Restoring the ecological relevance of water in people's lives opens up the possibility of being intimate with another kind of watery force that may dissipate fantasies of the normalized good life. In a highly modernized and constrained environment like Singapore and many other metropolitan areas, it requires patience and creativity to negotiate with a normalized mode of thinking and to forge a more ecological response to resource management. In practice, this might involve experimenting with new ways of relating to and encountering water, such as the partnership between mangroves and waters or the ABC Waters Program. At the very least, these provide a supplement to the dominant imaginary that distances water. It is equally important in such an effort that attention is paid to the ocean, which has often been backgrounded, seen as a resource and a water dump. The heavy water usage in nondomestic spheres must be foregrounded in water management and in broad discussions on a country's carbon emissions and environmental impacts. Taken together, all of these efforts might begin a process of doing water otherwise, of learning to speak a watery language and listening to water differently, and ultimately allowing water to seep into various relationships.

INTERLUDE: THE GENEALOGY
OF TAP WATER

Growing up in a city, drinking water smelling of chlorine
I always wish I had a story to tell that I am a child
of the Yangtze River or the Yellow Sea. Then I remember
how far I have been removed from these waters.
I listened to neighbours speaking
their ancestral rivers, lakes, oceans; some swear their pulses
still beat the rhythm of the sea.

I sigh, go home,
turn on the tap, make tea.
Let its warmth traverse my body
and ask:

Quiet, clean, civilised, what path
        have you meandered before arriving
                to irrigate me, what are your stories before
                        you travelled through the pipes?

                        The ones of you collected from dammed rivers
                        or falling from the sky, the ones of
                        you purified, free of
                        toxic traces, the ones of
                        you desalinated, carrying no more
                        saltiness of home, and the ones of
                        you measured, priced to

a wet commodity, ubiquitous
and shadowed.

But most of all, tell me of
the abundant worlds
    you nourish—

the cartography of urban
infrastructures you cradle and
cleanse along the way.
Concrete, asphalt,
cool steel pipes, it is my odd kin I taste
in the tea, brushing my lips
weaving the stories
of cities
into my body.

How I have taken you for granted
for so long. Might I call your name, tell
your watery memories, as I tell
the ocean's?
And proudly claim my fluid lineage:

We are children of tap water.[1]

5

THE SPROUTING FARMS

I started my journey into Singapore walking in a garden, a City in a Garden. Little did I know that inside this controlled and manicured environment, other things were sprouting—vegetable seeds were germinating, fruits were ripening, edible flowers gave colors to otherwise bland spaces. While many of the gardens are ornamental, even distinctively showy, others are becoming spaces of intense production. I had meals in various farm-to-table restaurants proudly featuring their locally sourced produce, something of a novelty in a country that imports 90 percent of its food. Yet, it was not until I met Darren Tan, the head of education of Comcrop farm, and he brought me to the roof of a shopping complex on Orchard Road (the renowned mall-lined street in Singapore), that I realized that here "local" means blocks away. The often-empty rooftop was filled with lively greens peeking out from rows of racks, bathing under the blazing tropical sun. Unlike the usual color composition of dark brown soil juxtaposed with green vegetables seen on farms, the hues here were cleaner and more metallic. The white multi-tiered racks were neat and sparkling, each row equipped with intricate-looking pumping devices. There was no soil in sight; instead, each plant was rooted in a square-shaped sponge sitting in a plastic net pot. Darren explained to me that Comcrop uses hydroponic growing methods, which means their produce is grown in a soilless environment, fed by a nutrient-rich solution. Opened in 2014,

Comcrop is Singapore's first commercial rooftop farm, providing produce including various types of herbs and some leafy greens to nearby hotels and restaurants.

In addition to the rooftop spaces, I later found that food farms occupied other city gaps: office buildings, schools, housing estates, hotels, and even inside shipping containers. Some of the many farms are the labor of Edible Garden City, a pioneer of urban farming in Singapore. As land for farming was too expensive to acquire in Singapore, when they started up in 2012, Edible Garden City utilized a range of unexpected and often neglected spaces: rooftops, community spaces, or gaps in existing ornamental gardens. Now part of its work focuses on transforming these gaps and cracks of commercial and residential spaces into food forestry. By the time I visited Citizen Farm, Edible Garden City's head office, and its urban farming division, I had already read a number of news articles featuring the farm, crediting it as the initiator of the urban food movement in Singapore. The farm is located in the middle of Queenstown, Singapore's oldest housing estate, and sits on the site of a former prison. Compared to Comcrop, Citizen Farm had a much stronger earthy hue. The farm was larger than I had imagined. Plants were not arranged neatly, some were grown in pots of various sizes and others in soil-beds, appearing to be at different stages of maturity. As I walked around, I began to realize that it was a large complex, comprising multiple farming sections. Some were located indoors. When I asked Darren Ho, the head farmer of Citizen Farm, about the seemingly less "organized" look of the farm, he laughed and explained that Citizen Farm hosts a community of farmers practicing a variety of growing methods, from outdoor soil farming to indoor aquaponic farming and cultivating mushrooms. The large building with a "do not enter" sign is an insect farm. Citizen Farm is also a community hub and conducts a range of workshops to support its grow-your-own-food movement. In short, it doesn't aim for a systematic and clean look, or a singular, streamlined production process. The produce is mostly distributed to restaurants. Locals can also become "citizens" by subscribing to receive its Citizen Box consisting of a variety of vegetables.

The surge of interest in urban farming has been linked to the sudden global food price spike in 2008. This included staple foods such as rice and wheat and triggered wide panic and unrest (United Nations 2011). There are also growing concerns over biosecurity due to the rising number of incidents of food contamination. Although some experimental ways of growing food and community farms have attracted noticeable media attention, the Singapore government has released a much more resolute and planned model. During a parliamentary speech in 2017, Koh Poh Koon, Singapore's senior minister of state for national development, announced that "farming will begin to resemble an industrialised production process, much like any other factory we have" (Hansard 2017). In a Facebook post, he (2017) wrote, "We cannot control the weather. But we can control how we want to manage the risks. I urge all our farmers to work together with government agencies to transform our farming sector into a more resilient and productive one." In the same year, the Agri-Food and Veterinary Authority (2017) unveiled the Farm Transformation Map, a highly technological and productivity-driven plan focusing on the prospect of intense factory-style farming that may take "quantum leaps in productivity" through integrated vertical and indoor systems, automation and robotics, operating on and occupying minimal human labour and space.[1]

At a time of rising environmental pollution, the damaging effects of climate change and related price fluctuations that have been deeply felt by the farming industry (which partially caused these environmental issues), technoscience and intensive farming/fishery and food labs are positioned by the Singapore government as the best way to prepare, mitigate, and create a safe and stable future foodscape (Hansard 2017). In 2019, the Agri-Food and Veterinary Authority (AVA) announced an ambitious goal to produce 30 percent of the country's nutritional needs by 2030. This entailed an increase from less than 10 percent, focusing on vegetables, eggs, and fish. Sky Greens, an automated nine-meter-tall multi-story vertical farm, is seen as an embodiment of the kind of intense farming that the state envisions. Panasonic Factory Solutions Asian Pacific, the Japanese electric giant's farming arm in Singapore, is another prominent player in the agrotechnological farming scene. Akin to its meticulously assembled electric appliances, its indoor farm environment controls temperature, humidity, and carbon dioxide. It also uses an artificial lighting system to support an intensive farming model and ensure stability (AVA 2017). By now, there are more than thirty indoor vertical farms in Singapore, up from six in 2016.[2] Locally produced food has become ever more attractive.

Although Singapore now imports most of its food, this hasn't always been the case. Until the 1980s, the country was self-sustaining in its supply, or close to it, in pigs, poultry, and eggs, despite limited land resources (Chou 2014). As noted in earlier chapters, the aggressive urbanization process and Singapore's river cleaning project have resulted in many farmers being forced to abandon their farms and move into public housing. Consequently, they needed to seek alternative livelihoods while farmland was converted to other development purposes. Only a small area was retained for farming along with some local coastal fisheries.

The newly released Farm Transformation Map introduces another round of land reform. The government announced that they would free up some land for high-tech farming and agricultural research and development. It would, however, concomitantly take back existing farming land for military purposes (Hansard 2015). Farmers who have persisted for decades must now compete with other new entrants in a bid for land in the new farming area. The AVA (2018) states that the successful companies in the first tranche of land tender will incorporate productive and innovative farming systems including automated greenhouses with smart control, data analytics to optimize growing conditions, and multi-story farms using LED lights, robotics, and soilless cultivation systems.

What to farm (or not) and how to farm has long been the subject of social, cultural, political, and environmental relations. In exploring the emerging, complex, and conflicted terrain of urban farming, I am interested in how the new wave of growing and managing food shapes and is shaped by our mode of living. This chapter is not a general review of urban agriculture or alternative food networks, which include but are not limited to rooftop farming, community gardening, guerrilla farming, and vertical farming, some of which do not exist in Singapore.[3] Rather, it focuses on the stories of a few small- to medium-scale urban farms in this intensely modernized urban space. How might a technological and productionist ethos shape the future of farming and reconfigure our relationship with the environment? As the state seeks to cast out "traditional" farming, which it deems inefficient, what might be lost in this process?

Provoked by these questions, this chapter explores the rise of urban farming through the politics of localism and the notion of care. In the first two sections, I examine and reveal the diverse forms of "local food" that have been evoked and deployed in urban farms. How has localism, in some contexts, been reduced to a narrow sense of geographic location? As the stories of various farms progress, I highlight that care is central to the thinking and making of localism. Drawing on feminist thinkers on care such as María Puig de la Bellacasa and Annemarie Mol, I examine multifaceted modes of care enacted in various farming practices and ask what is being cared for in the process of farming. The last section argues that by performing the work of thick localism, urban farming may enable us to make much needed room to forge a *situated care* that is rooted in the desire and labor to cultivate a wider web of life while grounding in a specific set of relations. In this context, farms are understood as practicing a mode of care that carries out social missions and experiments with and subverts the dominant imaginary of industrial farming. Situated care may offer a counterpoint to some universalizing and anthropocentric care narratives legitimized by an eco-modernist logic.

WHERE IS YOUR FOOD FROM?

To stroll in the supermarkets in Singapore is to witness globalization in full motion: broccoli from Australia, potatoes grown in Dutch soil, leafy vegetables harvested in Malaysia or Thailand; fruits of various seasons and climates mingle happily. Although I am conscious of Singapore as a tropical country, my body becomes less attuned to my locality given the omnipresence of air-conditioning while the abundant produce seems to whisper, *there are no spatial or temporal limitations; everything is possible and available.* In science and technology studies researcher Erika Amethyst Szymanski's account, the American supermarket promulgates "a placeless food culture and [contributes] to the estrangement of humans from their environments by selling the same plastic-wrapped prechopped broccoli season after season" (2018, 56). Here, vegetables are not sliced, but they are nevertheless trapped in plastic packaging. The roots of some leafy greens are set in sponges to preserve their freshness. Locating local produce in the supermarket is a test of one's patience. Once found, the majority seem to be eggs, prepacked salad leaves, a selection of leafy vegetables, and fish, mostly selling at a higher price than imported varieties. Labels such as "Freshly Picked," "Freshest," and "Locally Grown" are prominently displayed.

It is important to note that consumer demand for locally produced food varies in different urban contexts. For example, in some developed cities, local produce has helped to improve food accessibility and quality for low-income communities (McClintock 2010). In some developing countries, imported food is preferred, treated as a sign of progress. In the late 1980s Singapore, in preparation for a complete phasing-out of local pig farms, the city-state ran a five-week national campaign aiming to change people's perceptions so they would accept imported frozen pork (Centre for Liveable Cities 2018). The recent rise in popularity of local produce in highly developed and wealthy Singapore (and many other cities) in part hinges on fear of these same foods when imported.

In my conversations with some local farmers, they shared the view that an increasing focus on both health and the environment has contributed to the rising demand for local food in Singapore. In contrast to a sea of food from diverse sources that has traveled long distances, local produce in Singapore is heavily marketed as sustainably grown, pesticide free, and hence safer, fresher, and better for the environment.[4] Eating locally has become a growing trend, imbricated in a more sustainable lifestyle and a positioning that Anna Lavis, Emma-Jayne Abbots, and Luci Attala term as "eating-as-caring" for self-care, for loved ones, and caring for the environment (2016, 12).

In the traditional perception of localism, soil is understood as the connective tissue between human and the landscape. For example, the use of Singapore's own soil is highlighted in the design and construction of CapitaGreen, a certified green building that is said to represent the ethos of living in harmony with nature (see chapter 1). At the same time, this connection has frequently taken a range of problematic, nationalistic forms (Bauman 1992; Heise 2008). In agricultural contexts, localism has also been linked to the notion of terroir, or provenance—the place of production. In Singapore, the definition of local produce is loosely defined by the AVA as "food that is grown within your locality and in line with accepted good practices." One of the reasons that provenance or local food is highly regarded is in part due to the fact that "the ecological conditions implicated in production processes can be more easily discerned" (Morgan, Marsden, and Murdoch 2008, 12). Many of these same values are also frequently associated with urban farming, which is often positioned as being quintessentially local. Whether it is vertical farming, futuristic plant factories, or soil-based farming, one of the most appealing aspects of urban produce is the locality of the production, which not only suggests freshness (due to the proximity to supermarkets or restaurants) but also appears to offer safety as the provenance of the product can be traced.

And yet, the rise of vertical and/or controlled indoor farming and hydroponics takes on a very different tone from food provenance, or terroir, and challenges the notion of localism. Some rooftop or vertical farms are sun-lit, but many operate in completely controlled environments, from using LED lights as sunlight to a range of other introduced resources. In these farms, unmoored produce is not rooted in local soil and the specificity of the environment in a traditional way. Rather, the goal is explicitly and deliberately to *overcome* locale and seasonality for year-long harvest. Although the majority of the crops grown in these farms are leafy greens, microgreens, and herbs, a high-tech local farm, Sustenir Agriculture, whose controlled environment "allows them to produce vegetables that have no exposure to chemicals, pesticides, pollutants and even dirt," managed to produce a variety of nonnative crops including strawberries, an unusual food option for the local climate and environment (Singh 2016). The novelty of this "local" produce sparked positive responses among consumers, who did not seem to care in the least that the actual growing conditions have no association with the Singaporean environment.

My visits to various local farms revealed more nuances in this kind of high-tech urban localism. For example, I learned that seeds used on the farm are not necessarily sourced locally but from overseas. As one farmer explained to me, harvesting one's own seeds would add another three to four weeks onto the production cycle, as they have to wait for the plants to flower and then wait for the bees to come. The nutrients required for the plants are also very different at the stage of flowering and seed setting. It is understandable that start-up commercial farms need to optimize their productivity. Yet, it also demonstrates that the claim of locally grown can be an ambiguous statement, open for negotiation. To what extent is the produce grown locally? Which part of the farming process, seeding, growing or distribution, or labor, needs to be local?

Urban agriculture is also heavily promoted for environmental benefits, such as reducing its carbon footprint as a result of reduced transportation and packaging (Orsini et al. 2017). According to Dickson Despommier (2010), an environmental science professor who popularized vertical farming, current farming practices are detrimental to the environment, as they use large amounts of fresh water, deplete the soil, and are reliant on fossil fuel. In contrast, a vertical farm in an urban setting uses less water and fossil fuels, is capable of producing crops all year round, and is immune to weather-related crop failure. They therefore provide ultimate control of food safety (Despommier). Vertical indoor farming in Singapore and many other cities has been positioned as a form of technological solution that is suitable for space-conscious and high-density areas. Guided by the Farm Transformation Map, this focus on the efficiency and intensified productivity of high-tech farming (and reduced water footprint) is a particular feature of what local food is coming to mean in Singapore.

However, some researchers question the simplistic proposition of local production being equivalent to sustainability. For example, local produce may reduce food miles, but it has a high consumption of energy from using LED light to maximize yields in indoor farming (Goodman and Minner 2019; Engler and Krarti 2021). In a recent paper on controlled environment agriculture in New York City, Wylie Goodman and Jennifer Minner (2019) question the value and environmental benefit of this form of farming, given its high energy consumption and limited nutrients due to the small range of the produce. Some studies suggest that the distance food travels is relatively insignificant in terms of reducing its carbon footprint, in part because most of food's carbon footprint comes from on-farm production and food preparation in factories and kitchens (Greear 2016, 110; Avetisyan, Hertel, and Sampson 2014).[5] Jake Greear argues against a narrowed focus on carbon footprint analysis as being all too similar to the "narrowly economistic thinking that created the now globalized industrial food system and all of its problems" (2016, 111). Instead, Greear suggests that the discussion of localism needs to be situated in a web of environmental, socioeconomic, and political issues.

In urban geographer Nathan McClintock's account (2010), the increasing interest in urban agriculture, particularly in the Global North, can be attributed to an attempt to repair urban dwellers' alienation from their food sources and the natural environment caused by the development of capitalism and urbanization.[6] For Singapore, a country long dubbed an "air-conditioned nation" that relies on artificial cooling to overcome the local climate (George 2000) and provide comfort for its residents who live in high-rise housing and consume food grown hydroponically from sky-high racks, localism seems to have been reduced to mere geographical coordinates.[7] What kind of environment do consumers of these produce connect with? What connections are possible with these various types of unmoored farming? More importantly, what type of localism would a more grounded farming enact? With these questions in mind, I traveled to Quan Fa Organic farm, one of the rare certified organic producers and a rare open farm in Singapore.

I arrived early. Fabian, a second-generation farmer, was finishing up a farm tour for a group of disadvantaged children. Although the day was not much warmer than any of the others in Singapore, walking in an expansive open farm is a quite different encounter with heat in an otherwise air-conditioned country. I had earlier learned that Quan Fa's land lease was not extended by the AVA. Fabian was quick to bring the topic up when we met. He told me that their tender for the land in the newly dedicated farming area was rejected by the AVA. They were informed that their method of farming does not meet the requirement of productivity. The farm is likely to close when the current lease expires.

Compared to the hydroponic farms I had visited with a limited range of produce, Quan Fa grows a vast variety of vegetables and fruits, including choy-som, bok-choy, radish, bitter melon, spring onions, sweet potato leaves, chili, various types of beans, mangos, and much more. Fabian went on to say:

Crop rotation is our main farming technique. Right now, we are growing lady's finger on this patch. Next month once all the harvest is gone, we will change to another crop. This is to benefit the soil and protect the environment. Farming to us is not just farming food that is safe, it is also to keep the soil clean and nourished. Our way of farming will not be able to meet the factory style farming requirement. In order to meet that level of production scale, we will have to change our farming method. The type of produce will also be limited. For example, we need to grow beans to regenerate the soil. . . . We only plant the items that are suitable for this climate.

With the farm's imminent closure on his mind, Fabian seemed flat as we started the conversation, but he soon regained his smile and liveliness as we walked among the vegetable patches (see figure 5.1), a few of which

5.1 A small section of Quan Fa farm. *Source*: photo by author.

were grown just for fun. Fabian said he learned how to farm from his uncle who had been growing food and looking after the land since the 1990s. As we approached a field of sweet corn, he stopped and showed me some videos he took while the corn was flowering. "The sweet corns attract lots of bees. I am not a good photographer. If you were here last week, you would see how beautiful it was." But it was not just in the videos; bees and butterflies were buzzing and dancing around flowers of other plants as we were walking. I wonder about controlled-environment farming or farms that do not yield flowers so that it maintains shorter growth cycles and how they fail to provide a hospitable, reciprocal environment for pollinators and other nonhumans in the city. In Quan Fa, farming is a generational and more-than-human effort, labored and attended by bees, butterflies, soil, many invisible organisms, and humans.

I asked about the challenges the farm was facing in addition to the lack of endorsement from the AVA. Fabian told me that the increasingly stringent requirements preventing them from hiring foreign workers had become an issue, "We do weeding, and many other tasks manually. It is

labour intensive. . . . It is difficult to find locals who are willing to do this."
Fabian was also candid about the price of their vegetables that could be
higher due to the operational cost in Singapore: "The price in the wet market is a bit better. But supermarkets need to add their own margin. We also
supply our produce to lots of restaurants and retail customers directly."

Although the government calls on consumers to support local produce, farms like Quan Fa, whose method of growing represents a localism
that aims to "make our ecological relationships visible and accountable"
(Plumwood 2008, 140) are unwelcome. In her critique of bioregional-
ism, Plumwood (2008) warns against the danger of localism becoming an
atomism that disconnects us from broader environmental relationships.
Having labored in this area for thirty years and having held countless
education tours, Quan Fa and its generational knowledge of farming and
caring for soil is regarded as obsolete. In the case of the state's visioning of
future farming, the superiority of *local* pays little attention to the heritage
or ecological environment of the produce.

As I continued my exploration of Singapore's urban farms, I found
that under the banner of a reductive localism, other shoots were growing.
The Edible Garden City (Citizen Farm), mentioned in the introduction to
this chapter, demonstrates localism in other forms. Having started from a
community farm, a desire to reconnect is at the center of its model. Walking around Citizen Farm, I saw people of various ages farming together. I
learned from Darren that Edible Garden City does foodscaping, where they
work with various sites to design specific and situated farming practices.
But this farm space is created as an open space for urban residents to be in
touch with nature, soil and promote people to grow their own food. Darren said to me:

We also collaborate with specific organizations to work with people with
autism. . . . They will go through some training, then start to try out on the farm.
This is a serious social mission of ours . . . What gives us great joy is when a part
time trainee decides to be one of our full time farmers. We notice that their behavior, their personality change as they settle in. We also engage with the ex-convicts
to see how growing, nurturing and caring for plants may help them.

Edible Garden City also aims to address some thorny local environmental issues through farming. For example, it collects food waste from
restaurants and accepts waste donations from residents. These are then

composted along with other organic matter back into the soil. From soil-based and soilless farming, to opening up space for local community, and to incorporating local food waste into its growing loop, Edible Garden City is less specific on the exact modes of farming, yet their interlaced practice demonstrates a contextual, ongoing localism that "is thicker and more concrete than mere location" (Plumwood 2008, 144). This kind of thick localism involves working with and within the surroundings, and with other humans, attending to environmental issues in socially and culturally connected ways.

During my visit and discussion with local farmers, one issue that was highlighted was the high setup capital and expensive operating costs associated with farming in Singapore, which is often a factor in the price of the produce. In many parts of the world, local produce is at the heart of premium farm-to-table concept restaurants and plays a significant role in many high-end hotels that promote sustainable, fresh, and seasonal cuisines. The consumption of local produce has at times been criticized as contributing to food elitism or even leading to gentrification (West and Domingos 2012; Horst, McClintock, and Hoey 2017).[8] In this light, it is important to ask who benefits from urban farming. Who has access to this food? Who is excluded from the scope of localism? Furthermore, pricing is a key issue when considering vertical or other modes of high-tech urban farming methods. In their discussion of urban, indoor, and controlled farming, Kurt Benke and Bruce Tomkins point out that some kinds of produce are popular in vertical farming, not because of "any inherent limitations in crop types" but because they provide a "premium profit margin" (2017, 21).[9] The small to medium farms I visited mostly supply restaurants, hotels, and a small number of local families. For the larger-scale urban farms that supply supermarkets, the price is still not to their advantage due to high setup capital and expensive operating costs. Although the food may often be grown in marginalized (under-utilized) land, it may not be for people who are marginalized.

As I moved from one farm to another, it became clear that, in the context of urban farming, localism takes diverse forms and is subject to being enrolled, distorted, or redone. If localism is positioned to form part of a simpler, less industrial future, or represents an inextricable relation between produce and the surrounding environment, agrotechnological farming featuring unmoored food grown in a controlled environment may engender an ambiguous and narrowed localism. On the other hand, some farming methods are more attuned to the produce, soil, and ecological environment, yet their possible reliance on nonlocal labor may further complicate the issue of local. Localism can also be an ongoing process as it engages with local social issues, from tackling food waste to helping groups with autism. As we saw at Comcrop, Quan Fa, and Citizen Farm, these various contingent and highly constructed "locals" do different work either connecting or disconnecting, caring for different things, and ultimately making or unmaking the future food network.

In short, localism in urban farming is far from straightforward. Rather, it is an eco-entanglement of geographic location, the actors involved (humans and other-than-humans), and social relations, in which it is important to grasp what is local and for whom.[10] At the same time, an ontological discussion on localism is not sufficient to unravel how the many interconnected elements are held together or pulled apart. I will now focus on the notion of care to further my exploration of the multifarious terrain of urban farming. Thinking with and through care is central to this discussion of the local.

SITUATING CARE: WHO DO YOU CARE FOR?

As the dominant farming imaginary pushes for technological vertical solutions and indoor controlled intensive farming, a new vocabulary comes into operation. Both Darren and some other farmers I spoke to proudly and affectionately call themselves urban farmers. Sky Greens, seen by the state as the poster child of urban farming, attributes its success to engineering and technology, identifying itself as a solution-provider in agriculture.[11] The state refers to modern farmers as "agri-technologists" or "agri-specialists" and to farms as plant factories. These divergent exercises in naming in many ways shine some light on various forms of care enacted in farming: earthly, bodily, material, and technological.

For Quan Fa, composting and soil care is at the center of its growing method. As Fabian and I passed through a hill-sized dirt pile, he told me that it was their "care" in the making:

We primarily use vegetable waste for composting, mixed with a very small portion of brown waste, soil dust and coffee grounds. We don't add anything else to accelerate the process, or use machines to break them down with heat.

Things naturally break down, with the help of bacteria, fungi, and all these organisms in the soil. It just takes some time. . . . This pile will take about half a year to ripen. They actually last for quite a long time. We are doing less now as the farm will close soon. . . . We still have farms in Thailand and Malaysia. We will find a way to keep growing food that tastes good and is good for soils. . . . There is so much our soil can offer . . .

In Singapore, although practicing a sustainable farming model (such as pesticide-free or water-saving) is placed at the center of the new wave of urban farms, very few seek organic certification. Organic certification is difficult to attain. Until recently, Singapore did not even have a certifying body. As such, Quan Fa obtained their certification from Thailand after a lengthy and stringent process. The Thai certification also requires ongoing monitoring such as maintaining detailed records of seeding, harvesting, and crop rotation.

Sky Greens, the automatic multi-storied farm, demonstrates some of the complexity and contestations associated with organic certification in Singapore. Researcher Khoo Hong Meng previously argued that although Sky Greens composts and reuses organic waste, the vertical farm should not really be understood as organic:

Organic farmers leveraged organic matter in soil to feed a wide array of soil organisms that could interact positively with plant roots. According to the US Department of Agriculture (n.d.), an organic farm is one that demonstrates ability to integrate cultural, biological, and mechanical practices that foster cycling of resources, promote ecological balance, and conserve biodiversity. . . . It would be a challenge to incorporate diversity and balance ecology into the [SkyGreen's] A-Go-Gro towers. (2020, 7–8).

But Singapore has its way. In 2018, the city-state set up its own framework—the Singapore Standards for Organic Primary Produce, or SS 632, the world's first organic standard for produce grown in an urban and peri-urban environment (Singapore Standards Council 2019). In June 2019, Sky Greens became the very first local vertical farm in Singapore to be awarded the home organic certification.[12] The vegetables of Sky Greens are grown indoors in small plant boxes with soil-mix, grouped in a long, narrow trough elevated above the ground. In 2017, Sky Greens began to produce only mini vegetables to avoid the use of pesticides (Begum 2019). This means the vegetables at the farm are harvested when they are smaller, at between twenty-one and twenty-four days, as opposed to the usual forty days, before insects appear (Begum). Although Sky Greens does not use artificial lighting, SS 632 allows this farming method. Urban farms worldwide, including importers, exporters, and retailers can apply for Singapore's SS 632 organic certification.[13]

While there have been heated debates on whether organic farming is more sustainable for the planet (and much depends on the specific forms of farming and certification standards), my interest lies in the modes of care that are enacted in the way food is grown. It is crucial to investigate what might enrich or limit connections, and the way in which care is situated. If organic farming has been positioned as a particular kind of care to many people, Sky Greens, or Singapore's SS 632 in general, seems to represent and even encourage a selected, pragmatic version of urban organic farming. In the case of Quan Fa, the method of encouraging diverse crops, most of which go through a complete lifecycle grounded in a growing process that cares for and is cared for by other species, embodies a more relational care and thick localism. As Puig de la Bellacasa explains, care requires maintaining the possibility of a web of ecological relations "rather than only from their possible benefits to humans" (2015, 701). The endorsement of Sky Greens' method with its compressed production cycle designed to avoid pesticide use and its overall limited engagement with a wider ecological system (inherent in the practice of vertical farming) signals a problematic paradigm shift in framing and the governing of care in farming.

Further, Puig de la Bellacasa argues that the drive for care within the productionist model "has mostly been for crops *as commodifiable produce*" and as relatively isolated objects of care (2015, 700, italics in original). In an agricultural context, she proposes that to enact "interdependent care" is to grow not only food products but soil that cultivates a lively multispecies mode of life (706). Quan Fa's attendance to soil and their broader ethos of food production represents this kind of ecological care. Yet, in the dominant forms of farming emerging in Singapore, including a controlled farming environment and/or compressed growth cycle, there is an apparent lack of this kind of care for either soil or crop diversity. But in its place, there is *not simply* an absence of care. Rather, as we will see, there are novel means of care emerging here.

The work of feminist and science and technology studies has demonstrated that care is never innocent, but rather it is complex, contested, and "an ethically and politically charged *practice*" (Puig de la Bellacasa 2011, 90, italics in original; see also Fisher and Tronto 1990; Tronto 1993; Robinson 2011; Murphy 2015). In Chao's (2018) work on oil palm in West Papua, she unravels a multiplicity of ambivalent modes of care enacted in this unexpected space. In particular, she draws attention to the often-disregarded care work practiced by corporate actors and the scientists who are involved in the production of new varieties of oil palm. Drawing on Puig de la Bellacasa, Chao suggests a "nonidealized vision" of and even compromised care that "requires attending to conflicts and tensions inherent in the relational practices of knowing, thinking, and acting" (443).[14] In Singapore, it is important to ask, what kinds of compromised or contested care have been performed and by whom? How is Singapore's version of food security nonetheless a form of care?

Notably, the care underpinning Singapore's pursuit for food security and safety is more than merely positioning crops as commodifiable produce that primarily prioritizes profit (Puig de la Bellacasa 2015). Rather, what looms large is a paternalistic form of care grounded in decades of dominance and control over humans and natures as well as a taken for granted attitude of technological prowess and intensification as best practice. In the 2019 Global Food Security Index published by the Economist Intelligence Unit (2019), Singapore is in fact ranked as the most food-secure country in the world, scoring high on measures such as affordability, availability, and safety.[15] In this case, the persistent anxiety of the state over food security seems to be strongly associated with a professed care for its people. In Joan Tronto's body of work on ethics of care, she warns against paternalism as constituting the danger of care, suggesting the possibility of the development of profound inequity "when the caregivers' sense of importance, duty, career . . . are tied to their caring role" (1993, 170). In a comparable vein, Aryn Martin, Natasha Myers, and Ana Viseu point out, "Practices of care are always shot through with asymmetrical power relations: . . . Who has the power to define what counts as care and how it should be administered?" (2015, 627). Indeed, paternalist care then becomes a powerful rhetoric by the state to legitimize its actions.

As we have seen in this book, during Singapore's island-wide urbanization, this kind of care was implemented through demolishing the Kampongs, described as unruly and unhygienic, and relocating farms and backyard trade industries along the Singapore River, whose mode of living was identified as a source of pollution (Loh 2009; Tan, Lee, and Tan 2016). In the current wave of farm transformation, this strategy includes maximizing production levels and casting out forms of farming or types of produce that do not comply, paving the way to enforce a techno-scientific intensive way of farming. Although Singapore's pursuit for urban farming is not entirely driven by the productionist model of industrial farming critiqued by Puig de la Bellacasa, eco-modernist farming is nevertheless increasingly envisioned as factories in which plants are manufactured in a standardized, universalized way of growing life. Viewed in this light, they become a calculated resource while framed as "local" and "sustainable" produce.

Furthermore, a controlled and closed form of food production is proclaimed as providing increased security for the customer and benefiting the environment. A manager at a high-tech land-based vertical fish farm explained to me that farming fish in completely indoor controlled environments reduces the need to feed fish antibiotics. Their immune systems are therefore not compromised. Moreover, the productivity of land-based fish farms is not weather-dependent or impacted by surging environmental issues such as algal blooms or plankton that cause mass fish kills. In this way, he told me, "Consumers are better cared for and feel more relaxed as they have our safer products." In Singapore and around the world, land-based or offshore fishery has also increasingly been promoted as a more effective way to care for the environment in the sense that it does not pollute the sea as some traditional coastal fisheries may (Gunther 2018; Choy 2019).[16] The AVA has also started to develop closed containment aquaculture systems that "isolate fish from the external environment" as a way for coastal fish farms to shield themselves from environmental pollution (Centre for Liveable Cities 2018, 79). Geographer Benjamin Coles suggests that practices of consumption shaped by concerns about personal and family health, other people, and the environment, "'place' and 're-place' foods into a palatable imaginative geography of safe and, by extension, unsafe spaces and places" (2016, 154). Facing increasing uncertainties, these farming practices and the form of care enacted are animated by an intense "turning inward . . . as a response to an increasingly ugly and threatening public world" (Plumwood 2005, 4). This kind of inward-looking care may set up a dangerous model that turns its back on environmental issues that are present in compromised systems.

Reflecting on these multifaceted and sometimes troubled modes of care, some of which emerge from intensified urbanization and radical environmental change, I suggest that there is an increasing need to attend to a relational and *situated* care, rather than a form of bio-regionalist care rooted in isolation and attempt to separate from nature. Here I am drawing on Donna Haraway's (1988) discussion of situated knowledges that (in part) is about taking account of and being accountable for the particular relationalities that we make and inhabit. In this way, situated care is rooted in upholding the multiplicity of more-than-human relations while attending to partial ways of knowing and inhabiting the environment. The Edible Garden City model offers a glimpse of the work of situated care that may cultivate a healthy robust environment in the middle of the city. Darren explained that the Citizen Farm follows a technique of complementary planting to reduce pests. As well as traditional ways of composting, they use vermicomposting and black soldier fly composting (in collaboration with another group of urban farmers). This practice introduces worms to forge a transformative relationship with local food waste, food produce, and humans.[17] Further, a key thrust of Edible Garden City's ongoing commitment is not simply creating food forestry closer to communities but also its grow-your-own-food movement that aims to inspire people to find a space in a confined environment to be closer to farming, plants, and soil, and to encourage and actively push for participation not just on a specific day, or a weekend, but to continue this relational, experimental, and affective care practice throughout everyday life. In Darren's words, "We have been to urban farms in Japan, Taiwan, America. Lots of inspirations are from others. But this farm is designed for a country like Singapore." When we talked about the pressure of productivity and profitability, Darren said:

We are not the most productive ones among the farms. Profits are important. But how to get the profit, and what do we do with the profit is another thing. . . . In a large enterprise where it follows pure efficiency, one plus one always equals two. Here, it could be two, or three, or ten. There is room for kindness and collaboration and care. How have urban farms or grow-your-own-food placed social and environmental impact on people? . . . all these feelings need to be accounted for and put into consideration.

5.2 A section of Comcrop farm. *Source:* photo by author.

Darren's view on farming as relational care reminds me of Annemarie Mol's meditation on the continuity of care. In the context of medical science, she proposes the logic of care as an ongoing care that challenges other market-based mechanisms that have a beginning and an end for enacting care and responsibility (2008, 18). Mol writes, "Care is not a transaction in which something is exchanged (a product against a price); but an interaction in which the action goes back and forth" (18). Constituting an assemblage of general outreach, community gatherings, and workshops, the move-ment is a performative way of creating a network of local farming offering and extending a situated care that could intervene in the growing trend of regarding urban farming as a new mechanism of commodification.

Continuing my exploration of the agricultural carescape, I learned more at Comcrop, whose hydroponic rooftop follows a more automated approach (see figure 5.2). Darren shared with me his view on automation in farming and the essential care element that sustains it: "In the past, farming relies a lot on farmer's experiences. The present farmers use sensors and data track-ing for similar information and to monitor the growth of the plants, but in a more precise and timely way. . . . There has been lots of trial and error since the advent of our farm. This is a new area in Singapore. No one really has a script on how to grow, and what to grow. We test the water almost every day to ensure the right number of elements."

As I walked around the rooftop, sampling the herbs and some edible flowers, I noticed Darren would often run his fingers along the surface of the racks. "The devil is all in the details," he smiled and explained to me that pests and pathogens can be issues in hydroponics, and there is lots of dirty work behind this neat and futuristic looking rooftop farm: "The hygiene of the racks and the farm is paramount. A lot of work and our time is going into maintenance, to make sure the frames are sanitized. . . . Each of our frames has about 1000 plants. If there is any sign of pest, we must identify it immediately." As Mol argues, technology is not static or opposed to care, but rather it is an integral part of caring in the process of growing and working along with other elements. It is not only about being skillful but about "being attentive, inventive, persistent, and forgiving" (2008, 55). In Comcrop and other indoor-controlled farming using a more technological and automated approach, farmers tend to their produce's wellbeing through data tracking and real-time monitoring. This testing and experimenting invokes a distant, virtual type of care in conjunction with the physical care that Darren described as "the dirty work."

As nutrient solutions replace the complicated soil environment, as controlled environments are preferred for precision and stability, the plants, now living in a lonelier place without an expansive soil environment to help regulate their health, become much more dependent on the grower and their attentiveness and care, as well as being more demanding and sensitive in the ways they are cared for. As Mol insists, "Do not just pay attention to what technologies are supposed to do, but also to what they happen to do, even if this is unexpected" (2008, 49). I read Mol's proposition of technology here as an open, experimental, and relational mode of care. In the case of controlled farming, the use of technology for monitoring, tweaking, and adjusting becomes indispensable to the life and death of plants, engendering a form of "techno-intimacy" (Weston 2017). In other words, the attempt of cultivating a resilient, more automated farming practice that is disconnected from the environment in fact reimagines new forms of interdependency and care between plants (or animals) and growers, mediated through technology.

In the context of food production, anthropologist Kath Weston (2017) argues against the dichotomic understanding between a supposed intimacy associated with "face-to-face" human-animal relation on small farms and an alienated relation associated with "modern" farming. Rather, Weston maintains that it is crucial to attend to "the *conditions* of food production" (20, italics in original).[18] Indeed, what has become clear is that technology is not necessarily a cold and externalized mechanism but may function as an integrated part to situate care in producing and sustaining a certain mode of life. Approaching the emergent urban farming scape through the lens of situated care helps to move beyond a singular framing of a human-centered (enabled) mode of care, and to attend to more-than-human relations and their specificities. In this case, situated care also seeks to do the ongoing work that Eva Giraud and colleagues insist—to not only recognize more-than-human entanglement in world-making, and to further engage with the "obligations posed by particular relations" (2018, 74). Much more work is needed on the social and ecological implication of these new forms of growing and care-making.

From the bodily care for the environment, to an ongoing mode of care that may disrupt (to some extent) the entire commodification of urban farming as a process, to a kind of care that may encourage work in isolation, the carescape in farming is fraught with affect, tension, compromises, and the captivating promise of a healthier and more abundant food supply with less energy input. Care may also come with a high price tag. At the same time, as in many other parts of Singapore, farming and food production, in particular some of the environmentally friendly modes of farming that require more labor, are heavily reliant on the labor of foreign workers (Jalais 2021). This is of course not a singular issue in Singapore. Rather, the shadow care underpinning the making of a sustainable environment is rooted in the long history of the intertwined cheap care and labor inequity in the care economy (Gottlieb 2022; Tronto 2005).

Furthermore, a reduced version of localism to location embodied in various farms in part plays a role in allowing the promises of urban farming to be recruited by a techno-industrial-capitalist mode with its particular narrow forms of care. Aryn Martin and colleagues write succinctly, "Care is ambivalent, contextual, and relational" (2015, 631). Indeed, care has its limitations. As demonstrated in the various farms I visited, their farming practices cannot care for all, nor do they attempt to. When examining urban agriculture, it is essential to focus on the relations that the process of farming and their produce inhabit, sustain, open up, and close down. Ultimately, for urban farming to be a meaningful alternative to industrial farming, one that avoids a reductive localism, it needs to develop the sentiment of relational thinking and situating itself in ongoing care that does not stop with humans (many are focused on care for "consumers," while others have a broader sense of care for the community), but rather also includes the land, soil, and a wider environment.

WHAT FUTURE ARE YOU GROWING?

At the time I visited Comcrop, it was on track to open a much larger rooftop farming site in the suburb of Woodland (which has since opened). This enclosed greenhouse will have a much higher level of productivity, thanks to its "pest control, light and shade control and automated growing systems" (Lim and Barber 2018). This new greenhouse will have the capacity to provide produce to selective supermarkets. The ultimate aspiration of Comcrop, according to its founder, is to develop into a plant factory (Lim and Barber).[19] Meanwhile, Edible Garden City continues on its multi-pronged path, working with a range of organizations to expand the possibilities of growing food in unexpected parts of the city. It is holding more systematic workshops and events covering a wider range of topics on farming as care and continues to promote the grow-your-own-food movement in the urban context. They are also starting to explore more seriously the possible therapeutic benefits of urban farming. The mini vegetables of Sky Greens, now certified organic, are available at supermarkets with an even higher price tag. The vertical farm has also expanded its farming technologies export business. On the other hand, I was delighted to learn that Quan Fa managed to find a smaller parcel of land and may continue its farming operation when its current lease expires, although they will need to temporarily reduce the variety of their produce. The handful of other more traditional farms face an uncertain future beyond 2021 as their leases expire (Whitehead 2019).

The divergent paths these farms are taking provide examples of the complexity of urban farming. As some continue to offer hope by opening up ways of practicing farming as situated care, the question is, will others be enlisted as part of a more eco-modernist factory-style farming?

If some food movements such as slow food have been linked to the values of the past (Morgan et al. 2008), technoscientific food production gestures powerfully to the future.[20] Melinda Cooper reminds us that in anticipating a certain version of the future, our generalized alertness is transformed into "a real mobilizing force, compelling us to become the uncertain future we're most in thrall to" (2006, 125). In Singapore's urban development turned future-making project (Yuen 2011), the Farm Transformation Map, the land reformation, and a newly minted SS 632 organic framework have seen farming revived and swiftly integrated as part of an overarching imagining of a sustainable city—a future of agri-tech-capitalism that it has called forth.

However, to protect and care for the future of "certain forms of life" is to potentially "dispossess" others (Anderson 2010, 791). As intensive and indoor farming becomes more and more dominant as a form of urban farming, positioned as a superior choice for bio and food security, an emerging economic solution or a technological revolution to feed people (it is of course fundamental in many regions), other possibilities will be squeezed out in the process. For instance, Benke and Tomkins suggest that the prospect of fully automated urban farms may see funding for indoor-farming research compete with field hydrology and soil science (2017, 24). More recent studies on controlled farming have also warned that, as farming moves to indoor sterile conditions, there will be a significant loss of soil microbes of which some are crucial to plant and human health (Chiaranunt and White 2023).[21] Here, the imagery is of unmoored plants prevented from communication with soil as a medium. These plants cannot nurture a multitude of other organisms while the vision of fish traveling between floors through tubes manifests a troubling way in which we attempt to preempt and/or prepare for a certain future.[22] In the shadow of this future, a degraded environment is abandoned as much as possible to secure a safe, flourishing, and well-fed future for some humans.

As the new farming movement, in particular vertical controlled farming, is posited as a triumph of technological advancement, as light-assisted

farming modules pop up at various corners of the city ás a symbol of freshness, it might be easy to lose sight of the fact that the rise of these eco-
modernized urban farms is also the result of intense urbanization, soil
depletion, environmental pollution, population growth, and food injustice
in many parts of the world. In other words, it is a set of approaches to farming that have been erected on the damage that we have done to the earth,
a manifestation of disconnection from the environment, and for some,
an urge to reconnect. Furthermore, like many efforts to secure resources
from the impacts of climate change such as seawater desalination, these
new methods can be themselves energy intensive and thus contribute to
climate change with their unevenly distributed environmental impact and
driving another round of demand for energy. This kind of vicious cycle
needs to be considered here, too. Yet, in Singapore, despite the fact that its
current heavy dependence on imported food is the legacy of its eradication
of Kampongs and their associated farmland, urban farming is again seen
as an isolated mechanical activity; a bio-techno-infrastructure that can be
readily reintroduced as an anthropocentric farming scheme policy, land
control, and technology investment.

My intent here is not to dismiss urban farming; rather, I am genuinely
excited by the possibility of this mode of growing that may improve social
well-being and encourage the imagining of a city from a more-than-human
dimension. What I wish to highlight is the growing trend to celebrate a
certain version of eco-modernist farming future that is proclaimed as local
and sustainable, efficient, and stable without attending carefully to the
many socio-environmental issues. This situation asks us to think critically
about the concept of the local as it is being transformed, revigorated, or
incorporated in the age of globalization and climate change. Indeed, some
localism may become "atomism," as discussed earlier. That is to say, the
kind of localism and care-making that is not interested in restoring what
has been erased or repairing the rift ultimately promotes an attitude of isolation. Similarly, my ambivalence toward certain practices and narratives
of vertical farming, along with some other technological approaches and
their rhetoric of localism, is not that they are high-tech and therefore lack
care or warmth. Rather, my concern centers on the particular forms of care
that they enact. How might this emerging farming scape transform, shape,
and expand biocultural-social-technological relations within humans, and

between humans and nonhumans? How might we use this opportunity to forge a practice that is more attuned to environmental connection, locally and beyond?

As Darren from Citizen Farm explained succinctly, the rise of sustainable solutions in Singapore is great, yet there is a severe lacking in education in the labor of growing food, nutrition, food wastage, and consumerism. Without these discussions, in his account, "what these solutions are really for is not really clear. All these are why and how we design this space to be. We want to use different modes of farming to connect." Emerging from the cracks and gaps in an ultra-modern landscape, from garden to gardening, from a city in a garden to an edible city, from food waste to fertilizer, from living in a high-rise building to farming on the rooftop, the trajectory of farming in Singapore is neither a practice abstracted from the environment nor it is a planned unfolding. Here, the future and/or the connections the Edible Garden City and some urban farmers seek may not always be as precisely articulated or defined as the Farm Transformation Plan. They may pay less attention to controlling or transcending the environment in their practices, but there is more curiosity, patience, a strong level of commitment to be attuned to the environment and attend to specific human and other-than-human relations, and much room and desire to learn and perhaps be transformed. Their sense of uncertainty is accompanied by an ongoing situated care for the future of farming.

As urban farming is still an emerging area, there is a pressing need to intervene in the developing ethos of intensification that may reproduce a farm machine that refuses to recognize interdependency and interconnections. The future of urban farming practices must be grounded in efforts to decenter human exceptionalism, and in a more expansive, *situated care* that includes diverse humans and nonhumans while tending to their differences. Urban farming needs to be understood as not only attempts to overcome the constraints of space, or a new way to accumulate capital, but as a "political work" that is committed to focusing on existing issues rather than trying to find "an outside alternative" (Puig de la Ballacasa 2017, 11). In the current era of overlapping crises, cultivating a relational mode of care may help to acknowledge shared precarity, and to evoke responsibilities in an interdependent world.

Importantly, this chapter shows that the development of urban farming is not a dichotomy between the productive high-tech and traditional way of farming as the Farm Transformation Map seeks to establish. Instead, unmoored farms, thick localism, diverse modes of care, technological interdependency, and a preempted version of the future are all at stake. Growing in the heart of capitalism, urban farming in its diverse forms offers possibilities to make visible the labor used to sustain the skyscraper and city, to revive the continuity that we have with each other and the landscape. These emerging ways of inhabiting and caring for the environment help to reimagine an alternative that may subvert a capitalist way of pressing forward and to interrupt the industrial farming rhythm that has caused devastating loss of many kinds.

Of central interest to this chapter is an invite for a much more careful and patient examination of and a situated way of thinking about this movement of seeding, sprouting, and transforming. In engaging with or developing a new form of farming, it is equally essential to have clear definitions of such powerful terms as local, organic, and other expressions that matter to consumers (and producers). As we work our way through thinking and rethinking these terms and how they are embodied in practices, alternative ways of growing food and growing with food may open up, in and with an increasingly challenging environment. When growing things, it is important to ask how the process grows both us and a wider world.

EPILOGUE: A YEAR OF RECKONING?

On December 4, 2019, the Singapore government announced its final decision on the alignment of the Cross Island Line (CRL), the mass transit underground railway that is at the center of chapter 3. After years of consideration, the Ministry of Transport (2019) stated that the train line would take the shorter two-kilometer route, tunneling directly underneath the Central Catchment Nature Reserve (CCNR). In their account, according to the recently concluded phase II Environmental Impact Assessment (EIA), the disruption would be minor if all the necessary protections and mitigations are taken.[1] Furthermore, in the account of the Ministry of Transport, "In the longer term, it is a more environmentally-friendly option as the direct alignment has a lower energy consumption." Of course, it would also save the passenger six minutes of travel time (Ministry of Transport). The Nature Society Group (NSS) pointed out that the EIA has "acknowledged that residual impacts of works under and near the CCNR is still of 'Moderate' significance" (2019, 4). The NSS also emphasized their broader concerns relating to the possibility of unanticipated impact to the reserve during the construction. Nevertheless, they acknowledged the engagement work that the government had done. After all, compared to many other battles they have fought and lost, this time they managed to be involved at some level of consultation with authorities early on.

Although the state's decision on the direct alignment was not entirely unexpected, I was nevertheless stricken by the outcome of this long-debated issue. I felt deeply disappointed, followed by waves of anxiety. As Singapore has announced its grand ambition in underground development, would this become a landmark case that foreshadows or further emboldens many more projects to come? In many ways, the CRL, a human-centered and velocity-driven rail line envisioned as the future of sustainable movement and improved connectivity, exemplifies the complications in imagining and constructing a sustainable city in this troubled time.[2]

While I was trying to come to terms with the CRL alignment, the already months-long bushfires in Australia were raging, ravaging the land and its nonhuman and human inhabitants. The footage of a severely injured koala covered in blood frantically running across the highway was picked up by the media and broadcasted repeatedly. Yet, many more animals and plants were extinguished silently. The dire situation of the fires made people across the world gasp as Australia became the face of a country that indulged in fossil fuels and ignored the warnings of climate change (Krugman 2020). The image of children wearing protective masks due to hazardous air quality with a blood-red sky of Sydney, one of the world's most prosperous cities, as the backdrop, was dystopic and chilling.

Devastated and shocked as many were, little did we know that a perfect storm was brewing. The masks that were worn as shields from air pollution would soon be in desperate shortage globally due to an infectious disease, later known as COVID-19, a coronavirus pandemic declared by the World Health Organization (WHO). Although researchers are still investigating the exact origin of the virus, it is widely suggested that the zoonosis is linked to wildlife trading and consumption (Chakraborty and Maity 2020; Jiang and Wang 2022).

At the initial stage of the pandemic, Singapore was lauded as a role model for its efforts and success in preventing the spread of the disease, for its unparalleled ability to plan ahead and control, and a strong and paternal style of governing (WHO 2020).[3] Yet, its effort was soon hampered by the surge of infections among its migrant workers that would later catch the attention of the world and bring the city to a grinding halt. By April 2020, over 1,000 new cases a day were reported, out of which the vast majority were migrant workers.[4] Their cramped living conditions with poor ventilation, the shadowy side of the city, came to the surface (Yea 2020). Some of these workers might be the ones laboring in the Gardens by the Bay, suspended in the air maintaining an eco-modernizer's garden (chapter 1). Others will be building the Tengah Forest Town (chapter 2) or the CRL train line (chapter 3) to ensure the development of eco-living and sustainable movement. The city-state that takes so much pride in control was caught by surprise. Perhaps to them, these workers, who were placed outside of the country's social life, had always been invisible, just like the underground CRL, out of sight, out of mind. I wonder again whether the decision on the CRL, premised on engineering and scientific mitigations, may come back to surprise them in the near future.

Writing in the midst of these tragic and interrelated events, I am cautious of the need for "slow" scholarly work (Stengers 2018). Yet, as many have rightly pointed out, this pandemic is an issue long anticipated in part due to habitat loss, industrial farming, intense urbanization, capitalist modernity, and the resultant proximity of human and nonhuman and climate change (Anderson 2020; van Dooren 2020b; Gandy 2022a, 2022b). Some of these deep-rooted issues, such as overreliance on techno-capital intensive measures and social and environmental inequality, are issues that I have explored in this book, and more alarmingly, all in the context of enacting a sustainable and livable city.

What kind of sustainable futures are we calling forth amid the intensifying effects of climate change and geopolitical uncertainty, and at what and whose expense? What kinds of other possibilities are squeezed out in the process? How might we imagine differently? These are some of the questions that have guided my exploration and analysis in this research. They have also brought me to my central argument: the human-centered, high-tech, capitalist mode of urban development needs to be questioned and challenged. Doing so requires reimagining a more inclusive and diverse urban environment with a range of human and other-than-human elements, co-shaping cities and their social-cultural and economic relations.

As demonstrated in this book, attending to the city from a more-than-human lens does not sideline or flatten intra-human relations. Indeed, it is fully committed to inquiring into the shadowy places and unraveling the entanglement of social justice and environmental inequality.[5] The casting out of certain modes of life with the demolishing of the Kampongs and the replacement of "traditional" farming, as well as the drastically different treatment of urban wilds, from culling the unwanted ones to championing the others seen as the city's successful sustainable work, has made clear the politics of inclusion and exclusion at work. Fleshing out and holding onto these complexities is vital as it provides the foundation to attend to the particularity and multiplicity of these environmental issues. It is within these paradoxes and politics of urban sustainability that this book is situated and the three themes of eco-modernization, more-than-human relations, and future and futuring are negotiated.

As this book has moved between the past, present, and the imagined future of Singapore, it has drawn out the multifaceted and contested stories of urban sustainability. Exploring, collecting, and telling these stories is my attempt to weave together a more diverse, human and other-than-human, material and affective urban life. Engaging with the wider more-than-human world in reimagining urban sustainability expands the scope of urban planning. Not only does it reveal the profound disconnection and illogic of many taken-for-granted urban sustainability practices, thereby acting as an effective way to challenge business as usual approaches, but it also opens up space to move from the impasse of a narrowed velocity- and capital-charged urban sustainability model.

And so, let's trace the path this reimagining has taken, what and who have accompanied me along the way. The stories started from walking in a garden, a curated space that has long been seen as a bridge that connects nature and culture. In Singapore, the imagining and crafting of gardens through complex processes of greening and de-greening have been used to attract foreign capital. It is also a technology of control that enables the island's ambitions to discipline and govern the multispecies communities. The young country's acceleration into modernization and its persistent pursuit of development is inextricably tied to shifting authoritarian natures. Further, it shows how plants have been recruited into an eco-modernization narrative. For example, vertical/skyrise greenery is positioned as an exciting new approach to urban sustainability, yet at the same time these plants can become a replacement for trees and other ground greenery to satisfy the desire for continuous development. Importantly, I note that authoritarian nature never has full control as some humans and other-than-humans keep disrupting the mastery relationship. Foregrounding the garden as the main thread that ties the project of nation building, technology, and urban sustainability makes visible its powerful agency in shaping the city and the populace. It reveals how it facilitates and exerts control, and the synergetic formation of urban greenery and city. It also demonstrates that urban natures are never singular. Rather, the issue lies in which natures have been brought to the fore, to satisfy whose desire at what expense.

As is clear throughout the book, modernization and urbanization are also inextricably tied to temporal control. The normalization of modernist linear time directs a particular mode of futuring that disconnects and isolates the past. Its specific method of conservation and preservation also flattens the notion of heritage. Through the stories of the living Kampongs, the demolished forests, the exhumed tombs, the erected eco-housing, and the reimagined modern Kampongs, I challenge the legitimacy of displacing and relocating urban residents under an overarching narrative of development. When the unbecoming process of double erasure actively removes the built and natural environment, and memory and relations, it has a significant and recursive impact on the reduction of natural and cultural diversity. Furthermore, it fails to recognize that by augmenting (depleting) the present, it limits possible futures for many others. Here, the work of reimagining through a rich and layered more-than-human relation is also an effort of re-collecting and re-membering, resisting the singular ecomodernist's future.

Meanwhile, in urban environments, there has been a rise of repair work attending to the ecological connections that have been broken by past and ongoing transportation projects. As my research unfolded, it guided me to a wildlife bridge that seeks to rebuild the spatial and temporal connections of some animals. The usage of the bridge is not guaranteed, rather it relies on a contingent hope and the becoming of the bridge itself. This kind of openness and willingness to cultivate a multispecies movement is in stark contrast to the underground CRL, a project rooted in a singular belief in engineering and scientific mitigation efforts. These fleshy and entangled stories told by animals, the reserve and the bridge, environmentalists and the state make visible the shadowy side of sustainable transportation. Urban transport must be reimagined not just as a mechanical infrastructure project of human connection or comfort but within multiple modes of connection and movement, including their complex consequences.

If authoritarian natures in Singapore are underpinned by entangled mastery over natures and humans, the exploration of water infrastructure in the country reveals an eco-modernized future envisioning a decoupling from nature, albeit a necessarily illusory one. Seeing through the intensified development of seawater desalination and NEWater, the book illuminates the interdependence between technology, water, energy, high capital, and sustainable solutions. I argue that Singapore's approaches to shielding itself from a perceived and real vulnerability, including sea-level rise and resource scarcity, are underpinned by a meshwork of relentless capital accumulation and the exportation of environmental costs. The rise of technological water renders a perceived abundance of water, thanks to the celebratory technological breakthrough. Yet, the intensification of these kinds of water production obscures multi-scaled environmental issues and consequences. As we continue to satisfy the growing water consumption, an eco-modernized way of sustaining a perceived good life is ultimately premised on an illusory ecological invulnerability and a cruelly optimistic (Berlant 2011) sustainable future. What happens when the capital required to sustain a techno-green future dissipates? This is a question that seems to become less speculative and all the more urgent as Singapore and the rest of the world are entering economic recession due to COVID-19 and other ongoing geopolitical uncertainties.[6] Meanwhile, the various sites of the ABC Waters program and the restoration of urban mangrove forests enact their own rhythms, showing alternative ways of experiencing and living with water.

A central question of this research is how technological solutions might be positioned to shape a more inclusive and just future. In the examination and exploration of urban farming, technology as well as the farm products, the pollinators, and the soil have been foregrounded as more-than-human elements. The storylines of urban farming show a multiplicity of creative, troubling, or compromised modes of care: from plants or fish grown in vertical controlled environments to avoid pollution to Singapore's standard of organic farming that extracts and distorts many of the key aspects of building organic relations that support webs of more-than-human life. At the same time, the embodied care and labor that some urban farmers in Singapore have shown in attending to their produce compel and inspire us to rethink: What counts as good care, and what is being cared for in and through farming in urban locales? How has the notion of local at times facilitated a narrow imagination of rootedness? Thinking with this emerging practice in a developed, high-density urban environment, I advocate the need for a localism rooted in situated care. These passionate, invested ways of caring for the environment are a kind of reimagining that may ultimately break the pattern of intensification, externalizing, or decoupling.

As I write this conclusion, Singapore has announced that it will be moving from being a "City in a Garden" to become a "City in Nature" (Singapore Government, 2021). Since the advent of contemporary Singapore in the 1960s, the figuration of a garden underpinned by authoritarian nature has been central to the city-state's building toward a modern, efficient, unified green nation that attracts foreign capital. With the renewed force of eco-modernization, the decades-long City in a Garden movement is a systematic, techno-centric mode of naturing that has transformed Singapore into a particular kind of marketable urban sustainable model. The scheme reflects a paradoxical imagining of a city as being hospitable to nonhumans and at the same time relying on increasing controls and decoupling as a way to manage perceived risk (Wang 2021).

This book has captured some worrying patterns in the project, some of which are emerging and others that are new manifestations of deep-rooted issues of dualism that have long plagued urban policies and our lifestyles. For instance, the analysis of water and the rise of some technological natures points out that some humans seek to part ways with any elements of nature deemed harmful or uncontrollable through technology and wealth. Meanwhile, the reimagined *Kampong Kampus*, the ecological bridge, and the re-naturalized canals are welcome signs, showing a more experimental, inclusive, and relational way to urbanize.

Indeed, the book has questioned whether the human-centered, controlled garden is a helpful figuration in thinking of a sustainable, just, and heterogenous more-than-human city. Or borrowing what Lesley Head and her colleagues have asked, "If plants are understood as agents and subjects . . . how policies and governance would and should look different" (2014, 866). In this context, the proposition of a City in Nature might be a welcome one. At this stage, however, there are few details available to really allow us to understand what relationships and possibilities this new imaginary might produce. It seems that the key areas of the new proj-ect continue to focus on existing approaches, including extending park connectors and integrating nature into the built environment for a forest feel through skyrise greenery. The animals highlighted as part of this City in Nature are the same charismatic group, from butterflies to hornbills. The government has emphasized that many more gardens and parks will be built and abundant trees will be planted as it calls for communities to be stewards of nature. It is said that some nature-based solutions will be used to mitigate in-land flooding and sea-level rise due to climate change. Recalling the final decision of the CRL, the city-state's unwavering support for oil refineries, as well as the mode of authoritarian nature, I wonder whether the imagining of a city in nature is an enhancement of the same anthropocentric mode of naturing, a marketing strategy to further solid-ify Singapore's status of a livable city, or whether it reflects some genuine change in approach.

Although Alaimo has been concerned that "'sustainability' evokes an environmentalism without an environment, an ecology devoid of other living creatures" (2016, 176), cities around the world seem eager to make the urban environment more "natural." Yet, whether a City in a Garden or a City in Nature, these kinds of urban imaginaries are ultimately an invitation to humans to enter into a broader multispecies community, intentionally or unintentionally. In light of this, the question is how urban reimagining may genuinely cultivate the conditions for a more cohab-ited city. Indeed, a central argument and general concern of this book is that coexistence (multiplicity) is not a mere synchronization or a forced harmony, but rather it needs to be prepared for conflicts, to respect and sometimes live uncomfortably in an increasingly cohabited space. It needs to involve more discussions surrounding the "difficulties" of cohabitation and the compromises the city and many of its inhabitants need to make.

The sustainable city requires technological solutions, so too does it need a more nuanced and more-than-human understanding of sustainabil-ity and a willingness to experiment with and through the environment. Although the vision of City in Nature is yet to really differentiate itself from the City in a Garden approach, the city-state itself has shown some unex-pected and inspiring examples of spontaneous naturing. As coronavirus restrictions heavily interrupt the rhythm of some humans and of capital, and moreover as the foreign workers who maintain the city have been hard hit by the virus, the meticulously maintained greenery has reestablished some of its own rhythm (Asher 2020). There are general observations of the city becoming a lot more "wild." Some enthusiastic locals have reported sightings of not-often-seen flora and fauna. There have even been some discussions on whether Singapore should *allow* some mini-eco-systems to thrive through reducing the frequency of grass-cutting (Asher).

To borrow Marilyn Strathern's words, "it matters what ideas one uses to think other ideas (with)" (1992, 10). Reimagining a more-than-human city draws attention to and makes visible unexpected partners in the cosmos. It is also an apparatus of care, a work of expansion that engages us in exploring other possible futures. A critical part of the analysis and findings of this book includes problematizing and demonstrating the necessity to be clear about what is meant by powerful terms and concepts such as livability, local, connectivity, sustainability, care, and heritage in the face of intense urbanization and unevenly distributed environmental degradation. As we destabilize and challenge these taken-for-granted definitions in urban imagining, it elucidates other possible avenues for change.

Throughout the research, I have been guided and inspired by the efforts I observed and participated in during my time in Singapore and the many people I met along the way, from Kampong dwellers and conservationists to the researchers I spoke to on urban nature solutions and the volunteers who ran public educational tours aiming to generate care for natures. In my discussions with staff members from various government agencies on issues of urban greenery, green building, and overall urban policies, they show a clear sense of urgency and devotion in exploring a more engaged and open way of approaching environmental challenges in the city. Furthermore, whether it is the work of the Ground-Up Initiative to practice a slow and relational way of engaging with the environment and the initiatives of mangrove reforestation, or those of growing edible gardens in the cracks of capitalism, these reimagining works, although having their own compromises, explore and materially craft alternative possibilities.

If the bush fires in Australia were not sufficient in serving as a grave warning of what is to come, the COVID-19 pandemic that arguably has its root in environmental degradation and the considerable damage some of us have caused to nonhumans brings a direct blow to Singapore and the rest of the globe. Yet, akin to climate change, "we" are not experiencing or suffering from this latest crisis in the same way. This is evidenced by Singapore's mass COVID-19 infections among migrant workers and other marginalized communities around the world, from indigenous communities in Brazil to the unequally distributed impact of employment across gender, class, and race in America (Kantamneni 2020; Wallace 2020; Roy 2020).[7]

Nevertheless, as a result of a global lockdown to curb the spread of the virus, images of animals reclaiming the urban space around the world are emerging: from kangaroos and pelicans roaming in the middle of the city to swans waltzing in the canals of Venice (Lanzoni and Almond 2020; Daly 2020). Intriguingly, although these news articles and images had been widely shared as moments of hope, many were later revealed as hoaxes (Daly). What seems to be clear here is a yearning to be closer to some forms of nature in our cities. Like Singapore's unexpected natures that have delighted many, people *want* to imagine this future, and in imagining they help to make it possible and to advocate for it in some way. Might projects like a City in Nature take cues from these versions of reality, resisting the continuation of a singular narrative with a narrowed set of possibilities of life? Furthermore, it is recorded that air pollution levels have dropped significantly in China (McGrath 2020). Blue skies that had disappeared behind smog for years have reappeared in India (Slater 2020). Meanwhile, the oil market briefly crashed in part due to weak demand (Gaffen 2020).[8] Among the many calls in the midst of the pandemic, some are for a radical revision of humans' existing modes of living that have ravaged the environment. Others asked whether this could be a "portal" to something or somewhere else.[9] These events give a glimpse into another version of the urban and the world beyond, albeit through a devastating, unexpected, yet inevitable cause.

Yet, there are real risks that things may turn in other directions as the world seeks to recover from the pandemic. For example, there are already calls for more development of infrastructure projects to reboot the economy (IMF 2020). In particular, will Singapore's eco-techno authoritarian way of governing see a further centralization of power? Amid all these possibilities, we are reminded of the importance of a responsible reimagining work and the possibility of doing it. For example, facing the border closure of Malaysia and Singapore during the pandemic and later ongoing global logistic challenges, the urban farms I visited are embracing an increasing demand for their produce. It is critical that this surge of demand is not used as the catalyst for intense commodification on urban farming, recruited by a techno-industrial-capitalist mode of growing.[10] Rather as urban farming is regarded with renewed passion, how might this become a force to push for deeper and broader changes to our relations with the environment?

In Singapore, one of the wealthiest cities globally, control, development, and vulnerability, as well as sustainability and livability, take on new intensity or acquire entirely different meanings. Throughout this book, my hope has been to explore and discover how the imagination of a green city has come into being, whether it is a success or failure, and at what cost to whom. Urban stories cannot be told as a mechanical narrative or mapped out as planning discourse dominated by the rhetoric of certainty. Cities are always already more-than-human. The more we seek to flesh out and foreground its multiplicity of actors and desires, the closer we would be to reimagining a more diverse and inclusive future.[11]

Amid the pandemic, health ministers and experts in some countries have pleaded with the public, noting that the responsible and effective approach is to act as if you already have coronavirus. In their account, it is "to imagine you do have the virus and change your behaviour" (Medley 2020).[12] To borrow this injunction, I would like to conclude this book with another act of reimagining work, one that is admittedly speculative but full of real hope for possible change.

Imagine you are the trees living
in shallow pits, transplanted,
fallen—whenever and wherever needed.   And now reimagine
trees become woods,
seasons of memories, love
and life circulating.

Imagine you are a civet crossing
the road that separates, kills
earth. You tremble under cars
devouring fossil fuels at speed.   And now reimagine
infrastructures are portals connecting
spaces, times, lifeways.

Imagine you are the rivers after all rivers
are dammed, that you never sing with another river, your body
policed to service only some   And now reimagine
each drop from the tap is kin and it irrigates
your senses, that you (re) grow
an aqueous tail.

Imagine you are homes of the living and the dead,
now exhumed. Bones exposed. Memories, webs of life erased
to build another road another park—sustainable, but all it sustains
is one mode of life   And now reimagine
a storied place storying
seeds, animals, air and sand.

imagine a city of one rhythm
/one movement/one text/one kind of being   And now reimagine
a city of many worlds rooted
in care. Writing, memory, times
are multiple. You start the lines,
light, leaves, and otters inherit the story.

Imagine you are   now reimagining
there is no place, no being
to transfer the damage. The inhabitants
of a more-than-human city share not only
the past and the present, but also
the future.[13]

NOTES

INTRODUCTION

1. In 2011, a Siemens sponsored study, conducted by the Economist Intelligence Unit, ranked Singapore as the highest, well above average in its Asian Green City Index. Singapore is also rated as Asia's most sustainable city and fourth globally by the 2018 Sustainable Cities Index from Arcadis (2018).

2. Ministry of the Environment and Water Resources and Ministry of National Development will hereafter be cited within the text as (MEWR and MND).

3. Alongside cited materials, the account of urban nonhuman residents such as macaques and snakes as well as wildlife roadkill in this book draws on my interviews and conversations with the member of Animal Concerns Research and Education Society (ACRES), conservationists and primatologists in Singapore and local people.

4. For a more detailed discussion of human-macaque relation in Singapore, see Yeo and Neo (2010).

5. For example, see Chew Hui Min, "Errant macaque at Segar Road" (April 18, 2017); or Wong Pei Ting "Shocking that Singaporeans ask for wildlife to be 'returned' to the zoo, says Jane Goodall," (November 27, 2019). See also Sha et al. (2009). Although complaints about macaques have made the headlines of newspapers from time to time, tensions between human and macaques are rarely mentioned in any official narrative of a City in a Garden.

6. In addition, sterilization has become an increasingly common approach to address the issue deemed as "overpopulated" macaques. Although some have claimed it to be a more humane method, local primatologists explained to me that such methods severely interrupt the social life and in turn the survival of the affected macaques.

7. For insights into the population of macaques in Singapore, and the issue and implication of not being able to access accurate culling records, see Riley, Jayasri, and Gumert (2015).

8. Author interview with the staff member at the ACRES and local conservationists.

9. It is reported that the study on the contribution of the land-use change and forestry sector to carbon emissions was undertaken by the National Parks Board as part of preparations for Singapore's Biennial Update Report to the United Nations Framework Convention on Climate Change. See *Third Biennial Update Report* (National Environmental Agency 2018).

10. As noted by the Singapore government, emissions from international aviation and marine bunker fuels are excluded from Singapore's domestic figures as the city-island-state is an international air and sea transportation hub. See National Environment Agency (2018).

11. It is important to highlight that Formula 1 racing is itself a highly polluting, fossil fuel–charged event. In Ng's book on Singapore's energy consumption, he argues that the country's active development in the motor sports industry "is a setback for the environment, and runs counter to the energy conservation message" (2012, 220).

12. As urban planner Jane Jacobs pointed out long ago in her influential book *The Death and Life of Great American Cities*, "The notion that reek or fumes are to be combated by zoning and land-sorting classifications at all is ridiculous. The air doesn't know about zoning boundaries. Regulations specifically aimed at the smoke or the reek itself are to the point" (1961, 232). On the issue of air pollution in cities, for instance, see Graham (2015, 2016). Also see Timothy Choy's (2011) fascinating discussion on breathability, and the poetics and politics of urban air.

13. Despite the severe hazardous impact on the environment, the practice is continued annually due to its low cost and deemed efficiency.

14. Singapore seems to never cease fascinating people. In addition to the plethora of literature in the past decades on the city-state, Stephen Hamnett and Belinda Yuen's (2019) recent edited book has assembled a wide range of topics from urban planning perspectives providing some focus on what could happen next for the city as well as newly emerging issues due to climate change. For some interesting, recent texts on the topic of bioscience research from an anthropological and cultural studies perspective, see Ong (2016), of smart city, see Calder (2016); Kong and Woods (2018). BBC, *National Geographic*, and *South China Morning Post* have also over recent years paid attention to Singapore's urban development discussed in this book. This literature and media exposure has provided pertinent background information for my research. It is important to note that much of this literature endorses Singapore's style of central planning, mode of greening, and construction of a particular kind of sustainable narrative.

15. Although Paul Crutzen and Eugene Stoermer (2000) are not the first to propose the term Anthropocene, they have been credited with popularizing this discourse in the new millennium.

16. For some insightful works attending to the environmental issues of Singapore from the perspectives of conservation, animal geography, cultural imagining, and

contained natures, see Neo (2010); Barnard ed. (2014). In particular, the local environmental group Nature Society (Singapore) has published well-researched position papers on Singapore conservation issues and controversial development work. Drawing on some of these important works, this book seeks to broaden and deepen the discussions.

17. Unless otherwise noted, interviews in this book were conducted by the author between 2017 and 2018 in Singapore. In some cases, I have identified participants by name; in other cases, where more appropriate, I have referenced them anonymously or pseudonyms are used (and acknowledged in the notes section when they are first introduced in the text).

18. I am inspired by many great thinkers in this discussion; for some key texts engaging with this topic, see Soper (1995); Cronon ed. (1995a); Braun and Castree, eds. (1998); Wolch (2002); Haraway (2008).

19. O'Gorman's discussion is in reference to her proposed concept "imagined ecologies," exploring the world-making possibilities of knowledge and imaginings of nonhumans, and their consequences (2017, 3).

20. Joseph Huber and Martin Jänicke have been widely acknowledged as pioneers of the development of ecological modernization (see Spaargaren and Mol 1992; Hajer 1997). A few scholars see a distinction between the terms ecological modernization and eco-modernism, for example, on their positioning of the role of the state (Symons and Karlsson 2018). I see this distinction as relatively blurry and shifting. As such, in this book I have used these terms interchangeably.

21. First developed in a small group of European countries including Germany, the Netherlands, and the United Kingdom, ecological modernization has gained much traction in developing countries including China, Vietnam, and Singapore since the late twentieth century (Mol and Spaargaren 2000).

22. David Harvey is thinking with Maarten Hajer (1992) and Albert Weale (1992) on this point.

23. I will discuss in detail the concept of "cruel optimism" (Berlant) in chapter 4.

24. The first wave of ecological modernization theory placed a strong focus on technocratic development. It subsequently started to involve aspects of industry ecology and policy/economic instruments, for example, tax reform, see Andersen and Massa (2000).

25. Singapore is known for its high level of consumerism. Chua Beng Huat's (2003) *Life Is Not Complete Without Shopping—Consumption Culture in Singapore* remains a key text on this topic.

26. Researchers have also pointed out the simultaneous physical separation and the flow of capital between the location of an eco-city and its surrounding heavy industrial development zone and argued that green cities are becoming an alternative way of expanding industrial economies; see Caprotti (2014). For example, SSTEC lies within the broader context of the Binhai New Area, a new key development zone.

27. Val Plumwood highlights the danger of the neglected, denied, and "unrecognised, shadow places that provide our material and ecological support" from an

environmental justice perspective (2008, 139). My thinking on the inextricable relations between eco-modernization and social justice is very much inspired by this.

28. There are wide discussions and documentations on Lee's thinking of greening and its role in attracting foreign investment, capital, foreign talent, and tourists. Lee's (2000) book *From Third World to First* is a good source. In a 1965 speech for the opening of the underground car park and public garden at Raffles Place, Lee said that he "had visions of a beautiful garden, everybody passing by admiring the flowers and trees and so on. And at the same time, the garden would be bearing gold coins and silver coins underneath—invisible to all sight—as cars went whizzing by" (2).

29. In Christensen and Heise's tracing of the development of urban ecology, they point out that "political ecology emphasizes that there is nothing 'natural' about this process at all . . . we are urged to see the city as constructed by social, economic, and political forces." Thus, "Studies in political ecology that resist the 'naturalization' of urban processes, while enormously productive intellectually, have so far failed to have any discernible impact on policy, planning, and building in cities" (2017, 439). Bruce Braun has raised a similar point in his discussion of Chicago School Sociologists and other urbanists: their reliance on ecological concepts in urban studies in fact leaves "urban ecology entirely devoid of ecology" (2005, 636). For a detailed recounting of the development of urban ecology, see Ernstson and Sorlin, eds. (2019).

30. Jamie Lorimer's (2007) discussion on how "nonhuman charisma" is mobilized in the context of UK biodiversity conservation is helpful in understanding how urban wilds are heavily differentiated. Meanwhile, Wolch, Jason Byrne, and Joshua Newell (2014) point out the correlation between urban greening efforts and the subsequent gentrification of the area, and the importance of having a balanced approach in urban sustainability through careful action.

31. As Whatmore explains: "Such modes of enquiry neither presume that socio-material change is an exclusively human achievement nor exclude the 'human' from the stuff of fabrication" (2006, 604).

32. On the topic of living with "unloved others" and awkward flourishing, see a special edited collection, Ginn, Beisel, and Barua, eds. (2014).

33. In Deborah Bird Rose's work, she notes the recursive loop between life and death: "life is held in place because death is returned into place to emerge as more life" (2005, 124). However, *double death* breaks up recursivity: "The first death is ordinary death; the second death is destruction of the capacity of life to transform death into more life" (124).

34. The population size has become a contentious topic in Singapore, in particular since the release of the controversial Population White Paper 2013 that caused a rare protest in the city-state. See chapter 3 for detailed discussions. In the most recent planning document, the government seeks to emphasize the narrative of sustainable population size, sidelining the narrative of pursuing a targeted population size.

35. Singapore has strict control on its population mix. There is a strong distinction in policies between "foreign worker" and "foreign talent." See detailed discussion in chapter 1.

NOTES TO CHAPTER 1

36. In thinking with Ingold on walking as knowing, my use of the term "world-making" takes inspiration of Karen Barad's proposition of agential realism where "Making knowledge is not simply about making facts but about making worlds . . . about making specific worldly configurations" (2007, 91).

37. There is a rich body of scholarship on race, ethnicity, and migration in a range of contexts of Singapore. For instance, for multiculturalism/multiracialism as a social instrument, see Chua (2005); on multiracialism and national identity, see Ang and Stratton (1995); on ethnic representation in media, see Tan (2004); on topic on race and cosmopolitanism, see Poon (2009); on a queer analysis of urban development and migration, see Oswin (2019); and on migrant workers and urban development, see Wee, Lam, and Yeoh (2022).

CHAPTER 1

1. For more details of this fascinating eco-development project, see the official website of Gardens by the Bay, https://www.gardensbythebay.com.sg/.

2. Donna Haraway (1994) uses the term figurations to denote something that is both material and discursive. My use of figure here takes cue from her—a garden is a place, a "figuration."

3. I borrow the term fugitive seeds from Christian Brooks Keeve. Although Keeve's work is in a different context of plantationocene, and the entangled lives of seeds and Black diaspora, I am inspired by their thinking on "seeds are fugitive" and how this fugitivity undermines "the control and extraction of human and botanic labor."

4. Singapore City Gallery is an urban planning gallery. Its interactive and dynamic exhibitions showcase the transformation of Singapore since its independence and the city-state's future plans.

5. In chapters 2 and 4, I discuss in detail the cultural, socio-environmental implications of Singapore's island-wide urbanization, the displacement and relocations of multiple humans and nonhumans.

6. For full description of the image, see https://www.nas.gov.sg/archivesonline/photographs/record-details/61af7f03-1162-11e3-83d5-0050568939ad. Accessed via National Archives of Singapore.

7. As mentioned earlier, Singapore is ranked as one of the greenest cities globally in Asian Green City Index 2011. In 2018, Treepedia, an interactive map site created by MIT's Senseable Lab and World Economic Forum listed Singapore as one of the ten cities with the most percentage of canopy coverage based on aerial images, see https://www.weforum.org/agenda/2018/03/the-12-cities-with-the-most-trees-around-the-world.

8. Harvey Neo has written about contested natures in Singapore. For example, as the top criterion of the Singapore government when selecting conservation sites is that the site must be "natural and ecologically stable," citizens' interpretations of greening the city often fell short of the government's definition of natural.

9. To trace back further, Theodor W. Adorno and Max Horkheimer (1997) argued the bound-up dominance of nature and intra-human in *The Dialectic of Enlightenment*.

Plumwood refers to Adorno and Horkheimer in her discussion on "the treatment of nature in purely instrumental terms" (1993, 24).

10. As James C. Scott writes, "the garden is one of man's attempts to impose his own principles of order, utility, and beauty on nature" (1998, 92). Here, Scott is also drawing on Zygmunt Bauman's use of the garden in describing modernity.

11. Lee has been widely dubbed and remembered as the chief gardener of Singapore. For example, see Perks (2017).

12. In chapter 2, I give a fuller discussion on temporal entanglement of Singapore's greening practices including the heritage tree program that has been used to create a certain past.

13. Brenda Yeoh and Maria Soco suggest that a female foreign worker in Singapore becomes a "*de facto* careworker to tend to both children and the elderly" (2014, 173). Yet, although they serve in domestic space in Singapore, their existence in the society and rights are shunned through rigorous governance and regulation. Note although Yeoh and Soco refer specifically to careworkers in their research, no work permit holder is allowed to marry a Singaporean or permanent resident in or outside Singapore without approval from the Singapore government agency. Nor are they allowed to become pregnant. For details of Work Permit conditions for foreign workers, see Ministry of Manpower https://www.mom.gov.sg/.

14. As part of its population control scheme, in the 1980s, Singapore launched the "Graduate Mothers Scheme" campaign to discourage women with low education from having children while encouraging those in the more educated elite (often Chinese Singaporean) to have more. For discussion of this policy as an intersecting issue of gender, race, and class, see Heng and Devan (1995); Lyons-Lee (1998); Oswin (2010). Lucy Davis's (2011) work explores the excluding and culling of stray cats—seen as the unwanted in Singapore—along with SARS and the state's policy on population control.

15. In the context of oil palm plantations in West Papua and their relations with Indigenous Marind, Chao challenges us to rethink "violence as a multispecies act" (6). Chao argues that by valorizing more-than-human entanglement while not attending seriously to particular groups of humans and other-than-humans "subjected to the damaging effects of a proliferating plant" is itself a mode of anthropocentrism. Although Chao's work is in a different context, it is helpful to think with this important proposition to complicate the discussion of the multispecies urban environment.

16. The Green Mark Certification Scheme was first launched in 2005 as part of the city-state's effort to promote environment-friendly buildings. For details of the scheme, see https://www1.bca.gov.sg/buildsg/sustainability/green-mark-certification-scheme.

17. I learned the environmental attributes and design philosophy of CapitaGreen through my visit and exploration at the site. For more details on CapitaGreen, see *CapitaGreen: The Green Jewel of the Central Business District*, https://www.nparks.gov.sg/-/media/cuge/ebook/citygreen/cg12/cg12_capitagreen.pdf.

18. Tham's comment is featured in the article on CapitaGreen's opening.

19. Part of the scheme involves revamping the Singapore Botanic Gardens (SBG). Although the SBG is not a focus in this book, it is worth noting it briefly here. Founded in 1859, the SBG was a key site for the experimental cultivation of cash crops under the British colonial rule. Since Singapore's independence, the SBG and its function as a botany research center was heavily restricted as the primary role of garden shifted to assisting the modernization and urbanization required for city-scale greening. Today it is a place well loved by locals for leisure and exercise, and it is known for its orchid collections for tourists. Timothy Barnard (2016) has written about the SBG extensively. To the Singapore government, the Gardens by the Bay gesturing to the future is the anchor for Singapore's new garden regime. Notably, in 2015, Singapore Botanic Gardens was inscribed as a UNESCO World Heritage Site for its *cultural* landscape.

20. Both Park Connector Network and Nature Ways are promoted as the indicator of a sustainable green city. The network of park connectors will be doubled to 400 km by 2030 (MEWR and MND 2014).

21. The Landscaping for Urban Spaces and High-Rises (LUSH): Landscape Replacement Policy was first launched in 2009 and has since been enhanced multiple times to push for further incorporation of skyrise greenery, and in more recent years of urban rooftop farms. Note that this policy only covers development projects in certain areas (Urban Redevelopment Authority 2017). The replacement of the ground greenery with other kinds of greenery is at partial scale. Many secondary forests are continually sacrificed for new development projects without the need to be replaced.

22. This is part of the display information at the Singapore City Gallery.

23. For example, in recent research in Melbourne, Jacinda Dromgold and colleagues suggest that "the effectiveness of green roofs as invertebrate habitat is highly dependent on location" and the network of surrounding habitats (2020, 1).

24. The index was formally endorsed in 2010 at the 10th Conference of Parties to the Convention on Biological Diversity in Nagoya.

25. Singapore's version of biodiversity is very targeted and specific. More details on Singapore's conservation planning in chapter 2.

26. The Marina South duck ponds were paved over in 1992 as part of a general land reclamation for future development. In 2005, the government announced that the same land would be used as the site for Gardens by the Bay. It was also said that the pond presented a public health problem as a potential site for mosquito breeding, see Neo (2007, 195). As the risk of dengue fever is high in Southeast Asian countries, the possibility of mosquitos breeding can be a very effective way to legitimate decisions in the name of public safety (Neo).

27. In a comparable vein, O'Gorman contends that, in Australia and elsewhere, the framing of wetlands as natural places that are largely separated from humans and culture undermines "understandings of them as valuable to many people for a range of reasons" (2021, 5). This kind of separation and radically reduced framing is ultimately used to pit the conservations of certain places against some humans and other-than-humans (O'Gorman).

28. SEA STATE 9: proclamation Garden was on view at the National Gallery of Singapore from April 27 to October 27, 2019.

29. For an overview of the SEA STATE series, see the artist's website: https://www .seastate.sg/. SEASTATE and SEA STATE have been used interchangeably in various sources.

30. For details of the exhibition, see the exhibition catalogue "Ng Teng Fong Roof Garden Commission: Charles Lim Yi Yong," edited by Adele Tan (2019). To accompany the exhibition, Charles Lim and the team have created a series of podcasts on the topic of land reclamation featuring a diverse range of views from specialists in the fields of botany, geography, legal history, and more. See https://www.nationalgallery .sg/magazine/sea-state-9-proclamation-garden-podcast#Chapter4.

31. In Stoetzer's fascinating book, *Ruderal city*, she proposes the notion of "ruderal" as an analytical lens to trace and draw out the diverse human-nonhuman relations in and around Berlin, and "in the ruins of European nationalism and capitalism." For a multifaceted discussion attending to spontaneous forms of natures in urban environments, see Gandy (2022a).

32. Devas was speaking to *Today*, a Singaporean Newspaper, see Seah (2016). For an insightful opinion piece on *Planet Earth II* and the stories that the program does not include, see Hicks (March 15, 2017).

33. The content is paid and presented by Gardens by the Bay as part of the BBC Storyworks (n.d.), a commercial content studio.

34. Here, I draw on the experiences of my many visits to the conservatories over the years.

35. In Myers's insightful article on Gardens by the Bay, she suggests that the domes of Gardens by the Bay also offer "perfect respite from Singapore's seasonal choking haze" (2019, 133). My personal experience through multiple visits at different times of year does not always resonate with it, mainly because the haze is seasonal. However, Myers's point is worth highlighting in particular in the context of the deep entanglement of climate change, global capitalism, and the devastating impact of monocropping. And I agree with her point that the dome display performs a magic trick, as quoted previously in the text. My brief discussion on the haze comes in the later section of the chapter.

36. It is worth noting that Barnard (2016) suggests that Gardens by the Bay reflects Singapore's economic power control over nature in a spectacular horticultural display, much like the gardens of Versailles, pointing to the tension between manufactured nature and "nature." As discussed earlier, my discussion of natures focuses on which nature has been put to the fore and its complicated consequences in the context of eco-modernist mode of sustainable development.

37. According to Gardens by the Bay, these advanced technologies can help to achieve at least 30 percent savings in energy consumption, compared to conventional cooling technologies.

38. In the article, researchers have also highlighted the role of some significant developing economies such as China and India in global deforestation.

39. Thinking with Erik Swyngedouw, Myers suggests that the dome and what it supports emulates post-political "relations that hinge on a fundamental split that renders nature as object and resource subject to technological management" (2019, 142). Indeed, in focusing on the development of green cities, the discussions need to critically engage with the kind of post-political techno-future. As seen throughout the chapter, my discussion seeks to (re)center the ethos of authoritarianism in the discussion of Singapore's mode of naturing. The mutually reinforcing mastery of humans and natures and technology needs to be interrogated to discern forms of oppressions, particularly when considering the rise of eco-authoritarianism and the intensifying eco-modernist mode of gardening.

40. In my conversations with the NParks staff, they explained to me that the efforts of maintaining skyrise greenery varies, particularly if the building owners prefer a neat and clean look. The maintenance also has other challenges including safety. Detailed safety guidelines have been published by NParks on vertical green installations and maintenance.

41. Drawing on Isabelle Stengers, "involutionary momentum" describes a "reciprocal capture" that "binds plants and people in projects of co-becoming" (Hustak and Myers 2012; Myers 2017, 297). In this context, "plants and people are both in-the-making in sites like gardens" (Myers, 297).

INTERLUDE: WHO HOLDS MY NAME?

1. An earlier version of the poem was published at *Voice and Verse Poetry Magazine* (2019).

CHAPTER 2

1. Sophia and Sumit are pseudonyms.

2. On the topic of "the future" in other parts of Asia, see Bunnell, Gillen, and Ho (2018). For a fascinating discussion on visual narratives, architectural design, and urban futures in Asia, see Wong (2022).

3. For example, Adam points out that Japan and Russia proceeded to "'Westernize'" their social relations with time in the late nineteenth century (2002, 17). In particular she was drawing on the work of Nishimoto Ikuko (1997) and Manuel Castells (1996) on this point. See also Ogle (2015) for a broader discussion on plurality of times, globalization, nationalism, and modernism.

4. I borrow the notion "unproductive" from Adam, where she highlights that as the economic time value is tied to the creation of and globalized clock-time, any "cultural resistance to this norm is equated with backwardness," while any work that is not aligned with the capitalist value is being rendered as "unproductive" work (2006, 124). See also Adam (2002, 2004).

5. In Chua's full account on this point, "homeowners are inclined to be conservative and supportive of the status quo, in the interest of protecting their property value. . . . At election time, [the ruling government] pressures public housing residents to vote

for the ruling party, by threatening to withhold funds for upgrading the flats and the estate environment up to a level comparable with newer flats with improved amenities and services" (2014, 521). Other scholars have also engaged with the topics of public housing and nation building in Singapore, such as Goh (2005), Kong and Yeoh (2003) and Perry et al. (1997).

6. According to Chua (2005), the quota policy is supposedly fearing the reestablishment of Malay "enclaves." He further points out the "greater financial costs to the Malay and Indian households in this quota system." For example, "a Malay household who is interested in selling its housing unit will have to sell it to another Malay household, if the quota for the Chinese and Indians in the block is already filled. This would not only reduce the market size of potential buyers but also potentially lower selling price because of the generally relatively poorer financial conditions of the Malay population; often the Malay household is not able to accept offers from Chinese buyers who are able and willing to pay higher price for the flat" (8).

7. In Limin Hee and Giok Ling Ooi's (2003) work, they suggest that the stratified nature of HDB housing and the top-down style of town planning have replaced the more spontaneous and organic use of space in the traditional Kampong houses and caused the abstraction of space. Although primarily in the context of contemporary Kuala Lumpur, Malaysia, Tim Bunnell's (2002) work on the ways in which Kampongs are imagined and lived by various actors, and the spatial-temporal entanglement of Kampong, modernization, urban development, race, and ethnicity, is fascinating.

8. Anthropologists Vivienne Wee and Geoffrey Benjamin recorded the lives of once lively communities inhabiting various southern Islands through their field work. For some of the images of the island in 1982, captured by Benjamin, see *Island Nation*, a documentary project about the islands of southern Singapore https://islandnation .sg/story/serendipity-island/.

9. For more details of the Heritage Road and Heritage Tree program, see the NParks website https://www.nparks.gov.sg/.

10. Singapore has four Nature Reserves that are conserved under the Parks and Trees Act 2006. For details of these protected green patches, see the NParks website.

11. These issues on water are discussed in a fuller sense in chapter 4.

12. The Bukit Timah Nature Reserve in Singapore, one of the few remaining reserves that contains an original lowland forest, experienced serious side effects from overflowing visitors, referred to by locals as "love it to death" (Auger 2013, 48).

13. As mentioned earlier, Rose has developed the powerful concept "double death" within the ecological context across her work. Although my concept of double erasure is situated in very different context, I was inspired by Rose's mode of thinking.

14. Darren Koh, cited in Han (2015). On a broader discussion of the erosion of burial rituals and religious beliefs in contemporary Singapore, and specifically in the context of "Chinese community," see Kong and Yeoh (2003). In addition to the state's usual rhetoric on space-scarcity, Lily Kong and Brenda Yeoh explore how the narrative of cemeteries being the breeding ground for mosquitos is deployed to further intrude and shape "landscapes of death" in Singapore.

15. The number of Singapore cemeteries has been reduced from 113 in 1958 to 25 by 2005. De Koninck, Drolet, and Girard have produced some excellent maps focusing on the territorial changes of Singapore since the 1960s; see the text and image on the change of burial-scape (2008, 68–69).

16. These discussions bring to mind feminist Karen Barad's proposition of the inseparability of time, space, and matter—the iterative becoming of "spacetimemattering." As Barad puts it, "the past is never left behind . . . and the future is not what will come to be in an unfolding of the present moment, rather the past and the future are enfolded participants in matter's iterative becoming" (2007, 234; 2010).

17. Over the years, Tay Lai Hock, the founder of GUI, spoke about his vision through interviews with various sources. References to Tay's views in this chapter draw on these sources. See Interview with Tay Lai Hock (2012). Profiled by Lim Zi Song. *We Are Singapore*; Interview with Tay Lai Hock (2014). Interviewed by F. Toh. *Archifest*. Tay passed away in 2018 and was mourned by a wide community.

18. Chris is a pseudonym.

19. Bastian's (2012) fascinating proposition of turtle-clock fleshes out the incommensurabilities between the rapidly diminishing leatherback turtle population, an animal that has long been associated with steadiness, the lack of urgency in changing behavior that harms their existence (human shrimping technique), and the time lag in passing the corresponding conservation regulations.

20. See Tan Pin Pin's interview with *Cinematheque Quarterly*.

CHAPTER 3

An earlier version of this chapter was published as "Re-imagining Urban Movement in Singapore: At the Intersection of a Nature Reserve, Underground Railway and Eco-bridge" in *Cultural Studies Review* 25 (2), 2019.

1. See figure 16, "Cross Island Line: Discussion and Position Paper" (2013, July 18). Thanks to NSS for allowing me to use this map.

2. For details of the Love MacRitchie Movement, see https://lovemacritchie.wordpress.com/.

3. The final alignment of the CRL was announced in December 2019. See the epilogue for details.

4. This question was raised by Louis Ng Kok Kwang in Parliament. However, the minister for transport Khaw Boon Wan simply replied that "the catchment there is already served by the Circle Line and the upcoming Thomson-East Coast Line"; see Hansard (2016).

5. See Marx (1973).

6. This has now been revised to seventy meters following the final decision on the alignment in December 2019.

7. Tony O'Dempsey is a geographic information system (GIS) and remote sensing professional, a noted conservationist in Singapore, and a council member of NSS.

8. Here, Melo Zurita and colleagues are thinking with Jamie Lorimer's (2007) discussion on the role of charisma in biodiversity conservation.

9. Singapore has a high-level collaboration (exportation) with other countries in building eco-cities. C. P. Pow and Neo (2015) have written about some flagship eco-city projects including the Sino-Singapore Tianjin eco-city, from the perspective of "policy mobility."

10. Here, Tan and Huang were speaking to the *Straits Times*. The controversy of CRL has attracted much media attention, having also been reported on by *CNA*, *Today*, and more. These points were echoed in my conversations with local environmentalists. At the same time I was reminded by them of the complexity of the issue and the necessity of continuous engagement with policymakers in efforts to moving things to a more ecological direction. This often involves deep patience and ongoing negotiations.

11. The width and the shape of the Eco-Link@BKE are determined based on NPark's study trip to Switzerland and the Netherlands where eco-bridges have also been constructed (Chew and Pazos).

12. Road collisions between cars and animals often result in significant financial cost. For instance, it is estimated that each North American deer crash "costs society around $6,600, elk $17,500, and moose more than $30,000" (Goldfarb 2020); see Huijser et al. (2009). As the construction cost of wildlife crossings is high, the economic interest and potential savings are often highlighted in relation to the development of these green infrastructures.

13. France is credited to be one of the first that constructed an animal crossing, although the purpose of the crossing was for hunters to guide deer.

14. Although urban wildlife corridors come in a variety of forms, one of the key aims is to promote connectivity and movement for species. Researchers have pointed out that the functions of corridors can include being a habitat, conduit, or barrier (Douglas and Sadler 2011). In this chapter, I primarily focus on wildlife crossing as a connectivity technology.

15. The effectiveness of corridors is a complicated issue and still under debate. For some insightful discussions on biological corridors and theirs forms, functions, and efficacy, see Rosenberg, Noon, and Meslow (1997). For a roundtable discussion on the topic, see "Do Urban Green Corridors 'Work'?" *The Nature of Cities* (2015).

16. In the context of the Netherlands, Van Der Windt and Swart argue that "the success of the ecological corridor cannot be explained by its scientific soundness. Instead, its social robustness seems a better explanation." Further, they suggest that "The power of the ecological corridor concept is related to its vagueness, its flexibility and its metaphorical appeal" (2008, 130).

17. There is a body of literature suggesting that many considerations are required in designing and building a corridor to fit its purpose. In Alexandra Koelle's words: "these routes [biological corridors] that animals travel to get from one suitable habitat to another—for food, water, to mate, or for other reasons—are species-specific. A spider's habitat and its biological corridors exist on a different scale than a grizzly's, and provide for different needs" (2012, 652). See also Hardy et al. (2007).

18. Koelle is thinking with María Puig de la Bellacasa's work on care. As Puig de la Bellacasa argues, care cannot be an "empty normative stance disconnected from its critical signification of a laborious and devalued material doing" (2011, 95). Rather, care is always an affective and embodied labor and ethical obligation (2012). My discussion on various modes of care is in chapter 5.

19. Stephen Caffyn speaks to Maxine Chen (2017) from the *Mongabay*.

20. Around the world, green corridors are often presented as the indicator of a city's advancement in and commitment to sustainability and environmental protection measures, and increasingly an adaptation strategy in the age of climate change, see Hilty et al. (2006); Beier (2012). As mentioned in chapter 1, in Singapore, Park Connector Network and Nature Ways Program are also key green initiatives of a City in a Garden scheme. Nature Ways Program involves planting specific trees and shrubs along a route to facilitate the movement of some animals (NParks). These programs contribute to Singapore's image as the model of future green urban living. From my own experience in Singapore in 2018 with some Park Connector paths, I observed that the greenery along the pathway was often not dense (perhaps still at their early planting stage). The paths were not wide and became crowded easily. I frequently needed to give way to joggers or people cycling. Meanwhile, electric scooters sharing the same path traveled at high speed, which seemed a hazard to humans (in November 2019, Singapore banned all electric scooters from footpaths due to rising accidents involving the devices).

21. As reported by Chew Hui Min and Rebecca Pazos, from 1994 to 2014, an average of two a year critically endangered Sunda pangolins have been found dead on major roads around the two reserves. According to NParks, there were no dead pangolins found during April 2014 to October 2015. They believe it could be due to the bridge. No further data at the time of writing could be found. Sadly, there were more road kills including a pangolin near Mandai area, a new development site for an eco-park. A Mandai Wildlife Bridge, inspired by Eco-Link@BKE, was promoted as a mitigation method for the impacts of the development. I will discuss more on this issue shortly in a later section.

22. Further, Barter has noted that in Singapore, speed enforcement focuses primarily on "high-speed, high-capacity roads, rather than along ordinary arterials" (2013, 234).

23. I am inspired by Sarah Franklin's proposition of an ethical bypass, through her work in the context of biomedical health technologies. Franklin discusses a biomedical company that reprograms a normal body cell to perform "*as if it were an embryonic cell*" (2001, 346, italics in original). In doing so, it avoids political or religious objections to the use of human embryos for research, or "the production of replacement body parts from aborted foetal tissue" (346).

24. For more details on this "eco-project" and its potential impact on existing secondary forest and wildlife, see NSS (2016).

25. The EIA report was commissioned by Mandai Parks Holdings and prepared by Environmental Resources Management (S) Pte Ltd.

26. From a broad perspective, Cymene Howe and colleagues remind us that ecological driven technology is often the result of human degradation of environmental conditions (2016, 556).

27. For a fascinating discussion on anticipatory action and future geographies, see Anderson (2010).

28. Here, Colin McFarlane is thinking with Hinchliffe and colleagues (2005) who draw their inspiration from Isabelle Stenger's proposition of cosmopolitics.

CHAPTER 4

1. Zoë Sofoulis (2005) has some insightful discussion on modernity, large-scale engineering projects, and what she coins as "Big water."

2. There have been a number of sources on this topic over the years. For example, see Dali et al. (2014); Pak et al (2021).

3. The Ministry of the Environment and Water Resources was renamed the Ministry of Sustainability and the Environment (MSE) in 2020.

4. These assuring words were said by the minister of the environment and water resources, Vivian Balakrishnan, during a parliamentary address. They were then swiftly and widely circulated in the media.

5. For details on desalination plants, NEWater, or any water related information and policy in Singapore, the official website of the PUB is quite a useful source: https://www.pub.gov.sg.

6. For example, see Lee (2012); Tortajada, Joshi, and Biswas (2013); and Ng (2018).

7. Although the Middle East still holds the highest desalination capacity globally, there has been a sharp increase in Europe in the past decades over 1,600 percent (Eke et al. 2020). Joyner Eke and colleagues suggest that "The production capacities of plants in some parts of Europe and Americas are not as par with those in the Middle East because of rigid environmental protection laws" that are concerned with the environmental implications associated with full-scale desalination, from brine to greenhouse gas emissions and more (2020, 4).

8. In addition to Kaika, Swyngedouw, and Gandy's respective extensive discussions on commodified and domesticated water, there has been a robust scholarship exploring the entanglement of urban water, infrastructures, and sociocultural imaginings and how society and citizenship are co-shaped by water, water infrastructures, and technology. For example, see Linton and Budds (2014), Anand (2017); Loftus (2009); O'Reilly (2006).

9. Malaysia and Singapore have a history of tumultuous relations in terms of water provision. Former Singapore prime minister Lee Kuan Yew (2000) notes that Malaysia would threaten to cut off water supplies whenever there were differences between the two countries.

10. Worster's (1992) work *Rivers of Empire* is primarily in the context of the American West.

11. In response to the tension that flared up over water in 2018, a third desalination plant in Tuas opened swiftly on the 28th of June that same year.

12. Compared to the resistance to drinking treated water in some regions, the acceptance of NEWater in Singapore has been achieved by a combination of intense public campaigns including rebranding and authoritarian governance. See Timm and Deal (2018). According to the PUB, NEWater is only added to reservoirs during dry periods.

13. Singapore currently uses about 430 million gallons of water a day (PUB).

14. Interestingly, even while Hyflux is going through liquidation, some maintain their faith in it. Some locals told me, "Water security is the number one priority in the country, the government would never allow Hyflux to fall. They will step in if needed." As of 2020, Hyflux is still going through its painful process of liquidation. The government issued notice to take back Tuaspring Desalination Plant from Hyflux in 2019.

15. According to the Singapore's Climate Action Plan, improving energy efficiency remains the key strategy. See MEWR and MND (2016a).

16. To learn more stories on the impacts of the transformation of the river, see Christine Lim's (2014) novel *The River's Song*, published by Aurora Metro Books.

17. Recent research from Singaporean biologists suggests that critically endangered freshwater crabs living in the streams are at higher risk of extinction due to fragmentation and isolation of their habitats resulting from intense urbanization (Tay et al. 2018).

18. In a similar vein, Singaporean geographers Jun-Han Yeo and Neo (2010) demonstrate that some macaques in the city-state and their human neighbors have learned to adapt their behaviors to cultivate a shared human-macaque space.

19. See De Koninck, Drolet, and Girard's map that shows the expansion of Singapore's landmass through reclamation (2008, 25). Singapore's land reclamation not only alters and impacts its own landscape and modes of life but over the decades, it has imported enormous amounts of sand from its neighboring countries such as Indonesia, Malaysia, and Cambodia. Many of these countries have now made sand exporting illegal. There is growing attention and concern over the severe environmental consequences of sand dredging and damaging impacts on locals' lives in affected regions. See Kalyanee Mam's (2018) heart-wrenching essay. On the topic of sand mining and its deep and complex more-than-human impacts around the world, see Pilkey et al. (2022).

20. According to the 2019 IEA report, energy consumption in Southeast Asia has dramatically increased over the years. Without a change in policies, the energy demand in the region is set to grow by 60 percent by 2040.

21. As Astrida Neimanis points out, water can be further commodified and deterritorialized by some due to "scarcity and pollution" (2017, 20).

22. There is growing attention across disciplines to the various environmental impacts of desalination. For example, on the broad impact of desalination, as well

as specifically surrounding the issue of brine and its possible detrimental ecological impacts on the ocean and its environment, see Panagopoulos and Haralambous (2020). For a discussion on seawater desalination from a political ecology perspective, see Williams and Swyngedouw, eds. (2018).

23. Reported by the *Straits Times*, see Tan (2017).

24. Another important aspect to consider is where desalinated water is actually used. For example, G. L. Meerganz von Medeazza (2005) points out that in some regions where water scarcity has long been an issue, the practice itself becomes "ethically questionable" as desalinated water is used for a range of purposes such as tourism rather than serving basic domestic needs.

25. This is based on the assumption that the process will be strictly followed. More discussion on the issue of *certainty* can be found in chapter 3.

26. As mentioned earlier, I borrow the metaphor of engine from van Dooren (2017).

27. Cecilia Chen, Janine Macleod and Astrida Neimanis remind us that "In many cases, the achievement of domination over watercourses (however temporary) coincides with an intensification of social domination" (2013, 6). I have discussed mutual, reinforced control between humans and natures extensively in chapter 1.

28. In *The Coal Question*, William Stanley Jevons (1865) made the link between increased technological efficiency and the use of coal.

29. Wong has been working on coastal geomorphology, sea level rise, and climate change for the past four decades.

30. As seawalls tend to impede biodiversity in their vicinity, local scientists have been experimenting with various methods to increase the structural complexity of seawalls in a bid to improve species richness of intertidal communities; see Loke et al. (2017).

31. The Restore Ubin Mangroves project (R.U.M.) is an ongoing effort. For more details of the project and its regular community-based events, see their website https://rum-initiative.blogspot.com/.

32. Stengers's proposition of cosmopolitics (2005, 2010) highlights the necessity of engaging with multiple parties and forms of knowledge in connecting divergent worlds.

33. On multispecies stories of urban mangroves, see Judith (2020). For some insightful discussions on how mangroves may adjust to rising sea levels and shape the trajectory of a changing ecosystem, see Krauss et al. (2014).

34. The limited discussions on the ABC Waters program are mostly surrounding its effectiveness in stormwater management. For example, see Goh et al. (2017). For a broad discussion on how the program might be seen as a socioecological practice, see Liao (2019).

35. In these troubling times, the notion of hope and its ambivalence have attracted much attention among scholars. For instance, Eben Kirksey, Nicholas Shapiro, and Maria Brodine (2013) suggest that hope becomes thicker when it is grounded in specificity (be it a specific location or a species). van Dooren proposes that hope "grows and expands as we actively care for the future" in an uncertain world (2019, 209). In

telling the stories of the ABC Waters program and mangrove forest, and throughout the book, my use of hope is situated in some of these discussions and moreover led by specific reparative stories in an urban context.

36. The rise of projects such as the wetland mitigation bank are precisely why we can't treat this emerging hope as another creative way to continue business as usual; see (Robertson 2004). For an interesting discussion on building/growing an oyster reef as a living seawall and the heavy labor by humans and oysters underpinning the supposed nature-based solution, see Wakefield (2020).

INTERLUDE: THE GENEALOGY OF TAP WATER

1. This poem was first published in *Otherwise Magazine* (2022).

CHAPTER 5

The chapter is derived from the article "The Sprouting Farms: You Are What You Eat" published in *Humanities* 10 (1), 2021 Special Issue: Food Cultures & Critical Sustainability.

1. The Singapore Food Agency (SFA) replaced the Agri-Food and Veterinary Authority (AVA) in 2019.

2. The rapid increase of Singapore's urban farms has attracted much media attention. For example, see Geddie and Su (2019); Emont (2020).

3. There does not seem to be a definitive definition of urban agriculture. One definition I have seen often circulated is from Luc Mougeot: "UA is an industry located within (intra-urban) or on the fringe (peri-urban) of a town, a city or a metropolis, which grows or raises, processes and distributes a diversity of food and non-food products, (re-)using largely human and material resources, products and services found in and around that urban area, and in turn supplying human and material resources, products and services largely to that urban area" (2000, 11). It is important to note that urban farming is approached differently across regions. For a brief overview of urban agriculture in developed countries, see Mok et al. (2014); for discussions in developing countries, see Hamilton et al. (2014).

4. In a news article focusing on locally grown organic food, Chong Nyet Chin, the director of food quality and safety of Singapore's NTUC FairPrice supermarket, points out that expanding disposable income, increasing health concerns, and consumer awareness are some of the drivers of local consumers wanting to "go for green, go for local and go for niche" (Ong 2019).

5. Specifically, Misak Avetisyan and colleagues have focused on the greenhouse gas emissions "engendered during the production and transport of internationally traded food products" (2014, 417). Based on the findings, they suggest that local production does not necessarily reduce transportation emissions, but rather the focus should be on the technologies of production that contribute heavily to emissions.

6. In the context of urban farming in North America, McClintock points to urban agriculture's "multi-functionality, from its attempts to overcome disruptions in ecological cycles to its ability to reclaim public space, re-embed food production and consumption

with socio-cultural significance, and reconnect consumers with their food and the environment" (2010, 193). As mentioned, it is to pay attention to the specificity and situatedness of the form of farming practices and their associated benefits in different regions; for example, food banks don't really exist in Singapore and many other cities. See my emphasis on a relational and situated care in the later section of this chapter.

7. The obsession with an air-conditioned environment is also linked to the increased use of water and energy. For a discussion on how air-conditioning enables the extraction of labor in Singapore, see McNeill (2019). At the same time, it is crucial to remember that many kinds of work performed by migrant workers in the city-state are in the open in hot and humid weather conditions.

8. Urban agriculture and the notion of localism in food production varies across the regions and is associated with diverse kinds of imaginings. The cited texts here linking to very different contexts are some examples, which also show the complexity of the issue. For example, Harry West and Nuno Domingos write about the elitism of the Slow Food movement, while Horst et al. refer to the entanglement of urban agriculture and eco-gentrification.

9. On this point, Benke and Tomkins are also drawing on the research of Despommier (2010) and Ian Frazier's (2017) report that traces the origin of the concept of vertical farming.

10. My discussion of a thicker notion of the local (beyond location) is connected to an ongoing discussion in geography about how space is rendered meaningful as place. For further literature, see Tuan (1977) and Casey (1997).

11. For more details of Sky Greens, see https://www.skygreens.com/about-skygreens/.

12. The award ceremony was held on June 11, 2019, see Zulkifli (2019b).

13. See Oliver Moore (2019) for a very interesting discussion on vertical farming going organic. Intriguingly, in a 2018 assessment of the financial viability of high-tech agriculture (plant factories with artificial lighting) for growing leafy vegetables in Singapore, researchers find that consumers are mostly willing to pay a premium for safety and freshness, and specifically, for organic produce. Thus, the study suggests that "an organic label . . . can enable local producers to sell their produce at a higher price." And, "Given the debates on considering plants grown in artificial environments as organic, an immediate imperative will be to look into organic certification approaches" for hydroponics and other non-soil farming practices (Montesclaros, Liu, and Teng 2018). It will be interesting to follow the development of the organic certification in vertical farming globally.

14. See Puig de la Bellacasa (2012). For the entanglement of violence and care in the context of animal conservation, see van Dooren (2014, 2015).

15. The 2019 index also points out Singapore's vulnerability toward climate change–related issues such as sea-level rise. This may further tie to the government's insistence on planning for climate-resistant food production through technology and other measures.

16. During the interview, the manager suggested to me that the feeding method that some coastal fish farmers use could cause damage to the sea. However, he was reluctant to elaborate further on this point.

17. Edible Garden City works with Insecta on vemicomposting. For a discussion on vermicomposting as practices of care, see Abrahamsson and Bertoni (2014).

18. Weston's discussion explores the deployment of a techno-surveillance tool in tracking the animals that are to become food, and the unexpected intimacies created between humans and animals in the process. Although this chapter does not focus on animal farming, and engages with different kinds of tension, I seek to highlight Weston's point on the compelling need to focus on the *condition* in which food is produced.

19. Comcrop founder Allan Lim and CEO Peter Barber were interviewed by Tay Qiao Wei (2018) at JTC, a Singapore government agency.

20. Soilless production systems are discussed very critically by proponents of agro-ecology approaches as being unnatural. Some researchers counterargue and point out that agriculture itself is a humanmade activity. For an interesting debate on this, see Muller et al. (2017). Again, as I have argued in chapter 1, and in other sections throughout this book, artificial versus nature should not be placed at the fore of the discussion.

21. Very recently, the National University of Singapore launched the Research Centre on Sustainable Urban Farming. The center is said to focus on developing high-technology solutions in indoor urban farming—for instance, examining the plant-microbe-environment system in indoor farming as well as genome editing—to create some plants with traits "tailored for controlled environments." For details, see https://www.dbs.nus.edu.sg/surf/.

22. How hydroponic, vertical farming and indoor fisheries may interrupt the social life of plants and fish is not in the scope of this chapter. Nevertheless, there is some lovely literature on the topic of the lively modes of life of plants; for example, Michael Marder (2013) has written about the desire and livelihood of plants. For posthuman anthropology, see Kohn (2013). Anna Tsing (2015) has vividly revealed the complicated social life of the mushroom. Research has suggested that heavily cultivated plants have been "rendered 'deaf' and 'mute' through intensive agricultural practices and pesticide use" (Hustak and Myers 2012, 103; Paschold, Halitschke, and Baldwin 2006). But overall, there seems to be limited study on the cultural, social, and environmental impacts of commercial hydroponic farming or vertical indoor fisheries on human and other-than-human communities. Of course, this is in part because some of these practices are emerging.

EPILOGUE

1. The phase II EIA reports were also prepared by Environmental Resources Management (S) Pte Ltd (2019).

2. Soon after the announcement of the CRL, Joseph Chun (2020), a researcher in law at the National University of Singapore, brought attention to whether the current *Parks and Trees Act* covers the underground area. In his words, "how far under the surface does the designation and protection of a nature reserve extend to? Would the construction and operation of an MRT tunnel passing 70 metres under the surface of a nature reserve still be subject to the nature reserve provisions of the PTA?" Chun

suggests that the answer is not straightforward, particularly if it is considered in the context of Singapore's state power. What is clear is that serious attention needs to be paid to how nature reserves can be fully protected legally, specifically considering the impact of the city's increasing appetite in developing its subterranean space (Chun).

3. In February 2020, the WHO chief Tedros Adhanom Ghebreyesus and the WHO highly commended Singapore's rigor in managing the issue. In their tweet, it says, "Singapore is leaving no stone unturned, testing every case of influenza-like illness and pneumonia, so far, they have not found evidence of community transmission."

4. For more details of the COVID-19 cases in Singapore, see the Ministry of Health Singapore website https://www.moh.gov.sg/.

5. As I have argued throughout the book, more-than-human or multispecies studies do/should not flatten the complexities of intra-human relations. As van Dooren, Kirksey, and Münster put it: "Having escaped the tunnel vision of anthropos to the great world beyond, multispecies studies scholars are also working to carefully avoid a reductive, homogenizing conceptualization of human life" (2016, 14).

6. A number of global financial institutions, for example, the International Monetary Fund and the World Bank, have warned that the world will experience the deepest economic recession since the Second World War as a result of the pandemic and shutdown measures. See The World Bank (June 2020).

7. It must be noted that more comprehensive research is required in terms of the unevenly distributed and long-term impact of COVID-19.

8. It is important to keep asking questions and look for change amid the pandemic. For example, how might we think of cities differently as the world emerges from the pandemic? What can we learn from this dramatic period of global lockdowns? Milan is already thinking how they may reimagine the city including reallocating street space to encourage cycling and walking post lock down (Chatterton 2020). For a very recent discussion on the COVID-19 pandemic, extended urbanization, and possible pathways to a new, better mode of urban governance, see Ali, Connolly, and Keil (2022).

9. In writing about the Indian government's response to COVID-19, Arundhati Roy (2020) draws attention to the horrendous treatment toward domestic migrant workers in India and politically motivated hatred toward Muslims. She also warns against the risk of rising nationalism and a massive police state post pandemic. Finally, her essay asks whether the COVID-19 can be a portal "to imagine another world. And ready to fight for it."

10. At same time, perhaps unsurprisingly, in Singapore's Green plan 2030, described as a "whole-of-nation movement" by the Singapore Government, the development of high productivity and high technological farming has been featured as part of the goal of building a resilient future.

11. Although not the central focus of the book, it is worth noting that once highly promoted eco-cites have been plagued with updates on how they have become

ghost towns or have been scaled back to their original environmental commitments. These cities in various parts of the world, built on an imagining of the power of low carbon, sustainable technologies have been abandoned before they even fledged. See Sze (2015); Griffiths and Sovacoo (2020). Some ultimately became a simulacrum of the fossil-fueled present, see Günel (2019).

12. The infectious disease expert Graham Medley was featured in BBC Newsnight.

13. An earlier version of the poem was published as part of "Urban" in *A-Z of Shadow Places Concepts* (2021).

REFERENCES

Abrahamsson, Sebastian, and Filippo Bertoni. 2014. "Compost Politics: Experimenting with Togetherness in Vermicomposting." *Environmental Humanities* 4 (1): 125–148. https://doi.org/10.1215/22011919-3614962.

Abrahms, Briana, Sarah C. Sawyer, Neil R. Jordan, J. Weldon McNutt, Alan M. Wilson, and Justin S. Brashares. 2017. "Does Wildlife Resource Selection Accurately Inform Corridor Conservation?". *Journal of Applied Ecology* 54 (2): 412–422. https://doi.org/10.1111/1365-2664.12714.

Abram, David. 1997. *The Spell of the Sensuous: Perception and Language in a More-Than-Human World*. New York: Vintage Books.

Adam, Barbara. 1998. *Timescapes of Modernity: The Environment and Invisible Hazards*. London and New York: Routledge.

Adam, Barbara. 2002. "The Gendered Time Politics of Globalization: Of Shadowlands and Elusive Justice." *Feminist Review* 70 (1): 3–29. https://doi.org/10.1057/palgrave/fr/9400001.

Adam, Barbara. 2004. *Time*. Cambridge, UK: Polity Press.

Adam, Barbara. 2006. "Time." *Theory, Culture & Society*, 23 (2–3): 119–126. https://doi.org/10.1177/0263276406063779.

Adam, Barbara, and Chris Groves. 2007. *Future Matters: Action, Knowledge, Ethics*. Boston: Brill.

Adorno, Theodor W., and Max Horkheimer. 1997. *Dialectic of Enlightenment*. Translated by John Cumming. London: Verso.

Agri-Food and Veterinary Authority. 2017. "Future of farming." *AVA Vision* 1. Agri-Food and Veterinary Authority. Singapore.

Agri-Food and Veterinary Authority. 2018. "Steering the Future of farming." *AVA Vision* 1. Agri-Food and Veterinary Authority. Singapore.

Alaimo, Stacy. 2012. "States of Suspension: Trans-corporeality at Sea." *Interdisciplinary Studies in Literature and Environment* 19 (3): 476–493. https://doi.org/10.1093/isle/iss068.

Alaimo, Stacy. 2016. *Exposed: Environmental Politics and Pleasures in Posthuman Times.* Minnesota: University of Minnesota Press.

Alam, Ashraful, and Donna Houston. 2020. "Rethinking Care as Alternate Infrastructure." *Cities* 100:102662.

Ali, S. Harris, Creighton Connolly, and Roger Keil. 2022. *Pandemic Urbanism: Infectious Diseases on a Planet of Cities.* Cambridge, UK: Polity Press.

Alfian Sa'at. 2008. *A History of Amnesia.* (Rev. ed). Singapore: Ethos Books.

Allan, J. A. 1998. "Virtual Water: A Strategic Resource Global Solutions to Regional Deficits." *Ground Water* 36: 545–546.

Allen, John. 1991. *Biosphere 2: The Human Experiment.* Penguin Books.

Allon, Fiona. 2020. "Water Resources." *International Encyclopedia of Human Geography*, 233–240. https://doi.org/10.1016/b978-0-08-102295-5.10763-2.

Anand, Nikhil. 2017. *Hydraulic City: Water and the Infrastructures of Citizenship in Mumbai.* Durham, NC: Duke University Press.

Andersen, Mikael Skou, and Ilmo Massa. 2000. "Ecological Modernization—Origins, Dilemmas and Future Directions." *Journal of Environmental Policy & Planning* 2 (4): 337–345. https://doi.org/10.1080/714852820.

Anderson, Ben. 2010. "Preemption, Precaution, Preparedness: Anticipatory Action and Future Geographies." *Progress in Human Geography* 34 (6): 777–798. https://doi.org/10.1177/0309132510362600.

Anderson, Inger. 2020. "First Person: COVID-19 Is Not a Silver Lining for the Climate, Says UN Environment Chief." *United Nations.* https://news.un.org/en/story/2020/04/1061082.

Ang, Ien, and Jon Stratton. 1995 "The Singapore Way of Multiculturalism: Western Concepts/Asian Cultures." *Sojourn: Journal of Social Issues in Southeast Asia* 10 (1): 65–89.

Apostolopoulou, Elia. 2020. *Nature Swapped and Nature Lost: Biodiversity Offsetting, Urbanization and Social Justice.* Cham: Springer Nature Switzerland.

Appadurai, Arjun. 2013. *The Future as Cultural Fact: Essays on the Global Condition.* London: Verso.

Appel, Hannah, Nikhil Anand, and Akhil Gupta. 2018. "Temporality, Politics, and the Promise of Infrastructure." In *The Promise of Infrastructure*, edited by Nikhil Anand, Akhil Gupta, and Hannah Appel, 1–38. Durham, NC: Duke University Press.

Arcadis. 2018. *Citizen Centric Cities: The Sustainable Cities Index 2018.* Arcadis.

Asafu-Adjaye, John, Linus Blomqvist, Stewart Brand, Barry Brook, Ruth Defries, Erle Ellis, Christopher Foreman, et al. 2015. "An Ecomodernist Manifesto." Oakland: Breakthrough Institute. https://www.ecomodernism.org/manifesto-english/.

Asher, Saira. 2020. "Coronavirus in Singapore: The Garden City Learning to Love the Wild." BBC, June 14. https://www.bbc.com/news/world-asia-52960623.

Auger, Timothy. 2013. *Living in a Garden: The Greening of Singapore*. Singapore: Editions Didier Millet.

Avetisyan, Misak, Thomas Hertel, and Gregory Sampson. 2014. "Is Local Food More Environmentally Friendly? The GHG Emissions Impacts of Consuming Imported versus Domestically Produced Food." *Environmental and Resource Economics* 58 (3): 415–462. https://doi.org/10.1007/s10640-013-9706-3.

Bakker, Karen. 2012. "Water: Political, Biopolitical, Material." *Social Studies of Science* 42 (4): 616–623. https://doi.org/10.1177/0306312712441396.

Barad, Karen. 2007. *Meeting the Universe Halfway: Quantum Physics and the Entanglement of Matter and Meaning*. Durham, NC: Duke University Press.

Barad, Karen. 2010. "Quantum Entanglements and Hauntological Relations of Inheritance: Dis/continuities, SpaceTime Enfoldings, and Justice-to-Come." *Derrida Today* 3 (2): 240–268. https://doi.org/10.3366/E1754850010000813.

Barnard, Timothy P. 2014. "Introduction." In *Nature Contained: Environmental Histories of Singapore*, edited by Timothy P. Barnard, 1–8. Singapore: NUS Press.

Barnard, Timothy P. 2016. *Nature's Colony: Empire, Nation and Environment in the Singapore Botanic Gardens*. Singapore: NUS Press.

Barnard, Timothy P., and Corinne Heng. 2014. "A City in a Garden." In *Nature Contained: Environmental Histories of Singapore*, edited by Timothy P. Barnard, 281–306. Singapore: NUS Press.

Barr, Michael D. 2019. *Singapore: A Modern History*. London: I. B. Tauris.

Barter, Paul A. 2008. "Singapore's Urban Transport: Sustainability by Design or Necessity?" In *Spatial Planning for a Sustainable Singapore*, edited by Tai-Chee Wong, Belinda Yuen, and Charles Goldblum, 95–112. Dordrecht: Springer.

Barter, Paul A. 2013. "Singapore's Mobility Model: Time for an Update?" In *Megacity Mobility Culture: How Cities Move on in a Diverse World*, edited by Institute for Mobility Research, 225–242. Berlin and Heidelberg: Springer.

Bastian, Michelle. 2012. "Fatally Confused: Telling the Time in the Midst of Ecological Crises." *Environmental Philosophy* 9 (1): 23–48. https://doi.org/10.5840/envirophil2012913.

Bastian, Michelle. 2013. "Political Apologies and the Question of a 'Shared Time' in the Australian Context." *Theory, Culture & Society* 30 (5): 94–121. https://doi.org/10.1177/0263276413486679.

Baudrillard, Jean. 1993. "Hyperreal America." *Economy and Society* 22 (2): 243–252. https://doi.org/10.1080/03085149300000014.

Bauman, Zygmunt. 1989. *Modernity and the Holocaust*. Polity Press.

Bauman, Zygmunt. 1992. "Soil, Blood and Identity." *The Sociological Review* 40 (4): 675–701.

BBC. 2016. *Planet Earth II*. Television Series. London, UK: BBC One.

BBC StoryWorks. n.d. "The Forest That People Forgot." *BBC Studios*. https://www.bbc.com/storyworks/travel/garden-of-wonders/cloud-forest.

Begum, Shabana. 2019. "Vertical Farm Receives the World's First Urban Farm Certification for Organic Vegetables." *The Straits Times*, June 11. https://str.sg/o9EN.

Beier, Paul. 2012. "Conceptualizing and Designing Corridors for Climate Change." *Ecological Restoration* 30 (4): 312–319.

Benabou, Sarah. 2014. "Making Up for Lost Nature? A Critical Review of the International Development of Voluntary Biodiversity Offsets." *Environment and Society* 5 (1): 103–123. https://doi.org/10.3167/ares.2014.050107.

Benjamin, Walter. 1968. *Illuminations: Essays and Reflections*, edited by Hannah Arendt. Translated by Henry Zohn. New York: Schocken Books.

Benke, Kurt, and Bruce Tomkins. 2017. "Future Food-Production Systems: Vertical Farming and Controlled-Environment Agriculture." *Sustainability: Science, Practice and Policy* 13 (1): 13–26. https://doi.org/10.1080/15487733.2017.1394054.

Berlant, Lauren. 2011. *Cruel Optimism*. Durham, NC: Duke University Press.

Berlant, Lauren. 2016. "The Commons: Infrastructures for Troubling Times." *Environment and Planning D: Society and Space* 34 (3): 393–419. https://doi.org/10.1177/0263775816645989.

Biro, Andrew. 2013. "River-Adaptiveness in a Globalized World." In *Thinking with Water*, edited by Cecilia Chen, Janine MacLeod, and Astrida Neimanis, 166–184. London: McGill-Queen's University Press.

Blackburn, Kevin. 2013. "The 'Democratization' of Memories of Singapore's Past." *Bijdragen Tot de Taal-, Land- En Volkenkunde* 169 (4): 431–456. https://doi.org/10.1163/22134379-12340064.

Blomqvist, L., T. Nordhaus, and M. Shellenberger 2015. *Nature Unbound*. Report from the Breakthrough Institute.

Braun, Bruce. 2005. "Environmental Issues: Writing a More-Than-Human Urban Geography." *Progress in Human Geography* 29 (5): 635–650. https://doi.org/10.1191/0309132505ph574pr.

Braun, Bruce. 2015. "Futures: Imagining Socioecological Transformation—An Introduction." *Annals of the Association of American Geographers* 105 (2): 239–243. https://doi.org/10.1080/00045608.2014.1000893.

Braun, Bruce, and Noel Castree, eds. 1998. *Remaking Reality: Nature at the Millenium.* London: Routledge.

Brennan, Liam, Emily Chow, and Clayton Lamb. 2022. "Wildlife Overpass Structure Size, Distribution, Effectiveness, and Adherence to Expert Design Recommendations." *PeerJ* 10:e14371.

Brown, David. 1998. "Globalisation, Ethnicity and the Nation-State: The Case of Singapore." *Australian Journal of International Affairs* 52 (1): 35–46. https://doi.org/10.1080/10357719808445236.

Brydon, Anne. 1999. "A Moment's Notice: Time Politics across Cultures by Carol Greenhouse [Review]." *American Ethnologist* 26 (4): 993–994. https://doi.org/10.1525/ae.1999.26.4.993.

Bulkeley, Harriet, Vanesa Castán Broto, Mike Hodson, and Simon Marvin, eds. 2011. *Cities and Low Carbon Transitions.* London: Routledge.

Bunnell, Tim. 2002. "Kampung Rules: Landscape and the Contested Government of Urban(E) Malayness." *Urban Studies* 39 (9): 1685–1701. https://doi.org/10.1080/00420980220151727.

Bunnell, Tim, Jamie Gillen, and Elaine Lynn-Ee Ho. 2018. "The Prospect of Elsewhere: Engaging the Future through Aspirations in Asia." *Annals of the American Association of Geographers* 108 (1): 35–51. https://doi.org/10.1080/24694452.2017.1336424.

Butler, Judith. 2004. *Precarious Life: The Powers of Mourning and Violence.* London: Verso.

Butler, Judith. 2009. *Frames of War: When Is Life Grievable?* London: Verso.

Butler, Judith. 2012. "Precarity Talk: A Virtual Roundtable with Lauren Berlant, Judith Butler, Bojana Cvejić, Isabell Lorey, Jasbir Puar, and Ana Vujanović." *TDR: The Drama Review* 56 (4): 163–177.

Calder, Kent E. 2016. *Singapore: Smart City, Smart State.* Washington, DC: Brookings Institution Press.

CapitaLand. 2015. "A Vision Comes to Fruition." *CapitaLand*, October.

Caprotti, Federico. 2014. "Eco-Urbanism and the Eco-City, or, Denying the Right to the City?" *Antipode* 46 (5): 1285–1303. https://doi.org/10.1111/anti.12087.

Caprotti, Federico, Cecilia Springer, and Nichola Harmer. 2015. "'Eco' for Whom? Envisioning Eco-Urbanism in the Sino-Singapore Tianjin Eco-City, China." *International Journal of Urban and Regional Research* 39 (3): 495–517. https://doi.org/10.1111/1468-2427.12233.

Casey, Edward S. 1997. *The Fate of Place: A Philosophical History.* Berkeley: University of California Press.

Castelletta, Marjorie, Jean-Marc Thiollay, and Navjot S Sodhi. 2005. "The Effects of Extreme Forest Fragmentation on the Bird Community of Singapore Island." *Biological Conservation* 121: 135–155. https://doi.org/10.1016/j.biocon.2004.03.033.

Castells, Manuel. 1996. *The Rise of the Network Society.* Oxford: Blackwell.

Castree, Noel. 1995. "The Nature of Produced Nature: Materiality and Knowledge Construction in Marxism." *Antipode* 27 (1): 12–48. https://doi.org/10.1111/j.1467 -8330.1995.tb00260.x.

Centre for Liveable Cities. 2015. *Planning for Tourism: Creating a Vibrant Singapore.* Centre for Liveable Cities. Singapore.

Centre for Liveable Cities. 2018. *Food and the City: Overcoming Challenges for Food Security.* Centre for Liveable Cities. Singapore.

Centre for Liveable Cities and Civil Service College Singapore. 2014. *Liveable and Sustainable Cities: A Framework.* Civil Service College Singapore and Centre for Liveable Cities. Singapore.

Chakraborty, Indranil, and Prasenjit Maity. 2020. "COVID-19 Outbreak: Migration, Effects on Society, Global Environment and Prevention." *Science of the Total Environment* 728: 138882. https://doi.org/10.1016/j.scitotenv.2020.138882.

Chang, T. C., and Shirlena Huang. 2005. "Recreating Place, Replacing Memory: Creative Destruction at the Singapore River." *Asia Pacific Viewpoint* 46 (3): 267–280. https://doi.org/10.1111/j.1467-8373.2005.00285.x.

Chao, Sophie. 2018. "Seed Care in the Palm Oil Sector." *Environmental Humanities* 10 (2): 421–446. https://doi.org/10.1215/22011919-7156816.

Chao, Sophie. 2022. *In the Shadow of the Palms: More-Than-Human Becomings in West Papua.* Durham, NC: Duke University Press.

Chatterton, Paul. 2020. "Coronavirus: We're in a Real-Time Laboratory of a More Sustainable Urban Future." *The Conversation,* April 27, https://theconversation.com /coronavirus-were-in-a-real-time-laboratory-of-a-more-sustainable-urban-future-135712.

Chen, Cecilia, Janine MacLeod, and Astrida Neimanis. 2013. "Introduction: Toward a Hydrological Turn?" In *Thinking with Water,* edited by Cecilia Chen, Janine MacLeod, and Astrida Neimanis, 3–22. London: McGill-Queen's University Press.

Chen, Maxine. 2017. "How Effective Are Wildlife Corridors like Singapore's Eco-Link?" *Mongabay,* July 26. https://news.mongabay.com/2017/07/how-effective-are -wildlife-corridors-like-singapores-eco-link.

Cheong, So-Min, Brian Silliman, Poh Poh Wong, Bregje van Wesenbeeck, Choong-Ki Kim, and Greg Guannel. 2013. "Coastal Adaptation with Ecological Engineering." *Nature Climate Change* 3 (9): 787–791. https://doi.org/10.1038/nclimate1854.

Chew, Elaine. 2017. "Digging Deep into the Ownership of Underground Space-Recent Changes in Respect of Subterranean Land Use." *Singapore Journal of Legal Studies* (March): 1–17.

Chew, Hui Min. 2017. "Errant Macaque at Segar Road: Other Monkey Incidents and What to Do when You Encounter Wildlife." *The Straits Times,* April 18. https://str.sg /3QMs.

Chew, Hui-Min, and Rebecca Pazos. 2015. "Animals Crossing Eco-Link@BKE: Safe Passage for Creatures over Busy Highway." *The Straits Times*, December 11. https:// graphics.straitstimes.com/STI/STIMEDIA/Interactives/2015/11/feature-ecolink-BKE -national-parks/index.html.

Chiaranunt, Peerapol, and James F. White. 2023. "Plant Beneficial Bacteria and Their Potential Applications in Vertical Farming Systems." *Plants* 12 (2): 1–27. https://doi .org/10.3390/plants12020400.

Chong, Terence. 2010. "Introduction: The Role of Success in Singapore's National Identity." In *Management of Success: Singapore Revisited*, edited by Terence Chong, 1–20. Singapore: Institute of Southeast Asian Studies.

Chong, Fah Cheong. 2000. *First Generation*. Bronze. Singapore River. Sculpture.

Chou, Cynthia. 2014. "Agriculture and the End of Farming in Singapore." In *Nature Contained: Environmental Histories of Singapore*, edited by Timothy P. Barnard, 216–240. Singapore: NUS Press.

Chow, Alex. n.d. *Bidadari Cemetery. Singapore Infopedia*. Singapore: National Library Board.

Choy, Natalie. 2019. "ACE Reeling in Growth Opportunities with High-Tech Fish Farming." *The Business Times*, September 27.

Choy, Timothy. 2011. *Ecologies of Comparison: An Ethnography of Endangerment in Hong Kong*. Durham, NC: Duke University Press.

Christensen, Jon, and Ursula K. Heise. 2017. "Biocities: Urban Ecology and the Cultural Imagination." In *The Routledge Companion to the Environmental Humanities*, edited by Jon Christensen and Michelle Niemann, 436–445. London: Routledge. https://doi.org/10.4324/9781315766355-57.

Chua, Alvin, and Jan Yap. n.d. *Gardens by the Bay. Singapore Infopedia*. Singapore: National Library Board.

Chua, Beng Huat. 1994. *That Imagined Space: Nostalgia for the Kampung in Singapore*. Singapore: Dept. of Sociology, National University of Singapore.

Chua, Beng Huat. 1997. *Political Legitimacy and Housing: Stakeholding in Singapore*. London: Routledge.

Chua, Beng Huat. 2003. *Life Is Not Complete without Shopping: Consumption Culture in Singapore*. Singapore: Singapore University Press, National University of Singapore.

Chua, Beng Huat. 2005. *Taking Group Rights Seriously: Multiracialism in Singapore*. Working Paper No.124. Asia Research Centre. Murdoch University.

Chua, Beng Huat. 2011. "Singapore as Model: Planning Innovations, Knowledge Experts." In *Worlding Cities*, edited by Ananya Roy and Aihwa Ong, 29–54. Oxford, UK: Wiley-Blackwell.

Chua, Beng Huat. 2014. "Navigating between Limits: The Future of Public Housing in Singapore." *Housing Studies* 29 (4): 520–533. https://doi.org/10.1080/02673037 .2013.874548.

Chun, Joseph. 2020. "The Protection of Nature Reserves under the Parks and Trees Act—A Deep Dive." In *CML Working Paper Series (19/05)*. Singapore: NUS Asia-Pacific Centre for Environmental Law.

Çınar, Alev, and Thomas Bender. 2007. "The City: Experience, Imagination, and Place." In *Urban Imaginaries: Locating the Modern City*, edited by Alev Cinar and Thomas Bender, xi–xxvi. Minneapolis and London: University of Minnesota Press.

Clark, Graeme, Nathan A. Knott, Brett M. Miller, Brendan P. Kelaher, Melinda A. Coleman, Shinjiro Ushiama, and Emma L. Johnston. 2018. "First Large-Scale Ecological Impact Study of Desalination Outfall Reveals Trade-offs in Effects of Hypersalinity and Hydrodynamics." *Water Research* 145 (November): 757–768. https://doi.org/10.1016/j.watres.2018.08.071.

Clark, Graeme, and Emma Johnston. 2018. "Desal Plants Might Do Less Damage to Marine Environments than We Thought." *The Conversation*, September 20. https://theconversation.com/desal-plants-might-do-less-damage-to-marine-environments-than-we-thought-103593.

Clark, Tanya, Tara Rava Zolnikov, and Frances Furio. 2021. "Wildlife Corridors: Urban Wildlife Corridors and Their Multiple Benefits." In *The Palgrave Encyclopedia of Urban and Regional Futures*, edited by Robert Brears (Editor-in-Chief), 1–4. Cham: Springer International Publishing.

Clements, Claire (Director). 2016. *Wild City*. Television Series. Beach House Pictures.

Clevenger, Anthony P., Adam T. Ford, and Michael A. Sawaya. 2009. *Banff Wildlife Crossings Project: Integrating Science and Education in Restoring Population Connectivity across Transportation Corridors*. Final report to Parks Canada Agency, Radium Hot Springs, British Columbia, Canada.

Clevenger, Anthony P., and Marcel P. Huijser. 2011. *Wildlife Crossing Structure Handbook Design and Evaluation in North America*. FHWA-CFL/TD-11-003. Lakewood, CO: U.S. Department of Transportation.

Cohen, Maurie J. 1999. "Sustainable Development and Ecological Modernisation: National Capacity for Rigorous Environmental Reform." In *Environmental Policy and Societal Aims*, edited by Denis Requier-Desjardins, Clive Spash, and Jan van der Straaten, 103–128. Netherlands: Springer. https://doi.org/10.1007/978-94-011-4521-3_5.

Coles, Benjamin. 2016. "Placing Security: Food, Geographical Knowledge(s) and the Reproduction of Place(less-ness)." In *Careful Eating: Bodies, Food and Care*, edited by Emma-Jayne Abbots, Anna Lavis, and Luci Attala, 151–172. London and New York: Routledge.

Collard, Rosemary-Claire, Jessica Dempsey, and Juanita Sundberg. 2015a. "A Manifesto for Abundant Futures." *Annals of the Association of American Geographers* 105 (2): 322–330. https://doi.org/10.1080/00045608.2014.973007.

Collard, Rosemary-Claire, Jessica Dempsey, and Juanita Sundberg. 2015b. "The Moderns' Amnesia in Two Registers." *Environmental Humanities* 7 (1): 227–232. https://doi.org/10.1215/22011919-3616425.

Conley, Verena Andermatt. 2012. *Spatial Ecologies: Urban Sites, State and World-Space in French Cultural Theory.* Liverpool: Liverpool University Press.

Connerton, Paul. 2008. "Seven Types of Forgetting." *Memory Studies* 1 (1): 59–71. https://doi.org/10.1177/1750698007083889.

Connolly, Creighton, and Hamzah Muzaini. 2022. "Urbanizing Islands: A Critical History of Singapore's Offshore Islands." *Environment and Planning E: Nature and Space* 5 (4): 2172–2192. https://doi.org/10.1177/25148486211051082.

Cooper, Melinda. 2006. "Pre-Empting Emergence: The Biological Turn in the War on Terror." *Theory, Culture & Society* 23 (4): 113–135. https://doi.org/10.1177/0263276406065121.

Cossé, Vincent. 2011. "A Practical Approach: Incentives for Skyrise Greenery." *CITYGREEN* 2. Centre for Urban Greenery and Ecology: 10–11.

Cresswell, Tim. 2010. "Towards a Politics of Mobility." *Environment and Planning D: Society and Space* 28 (1): 17–31. https://doi.org/10.1068/d11407.

Cresswell, Tim. 2014. "Mobilities III: Moving on. Progress in Human Geography." *Progress in Human Geography* 38 (5): 712–721. https://doi.org/10.1177/0309132514530316.

Cronon, William, ed. 1995a. *Uncommon Ground: Toward Reinventing Nature.* New York and London: W.W. Norton & Company.

Cronon, William. 1995b. "The Trouble with Wilderness: Or, Getting Back to the Wrong Nature." *Environmental History* 1 (1): 7–28. https://doi.org/10.2307/3985059.

Crutzen, Paul, and Eugene Stoermer. 2000. "The Anthropocene." *IGBP Global Change Newsletter* 41:17–18.

CSIRO. 2017. *What Does Science Tell Us about Fugitive Methane Emissions from Unconventional Gas?* CSIRO, Australia.

Cuff, Dana, Anastasia Loukaitou-Sideris, Todd Presner, ZubiaurreMaría, and Jonathan Jae-an Crisman. 2020. *Urban Humanities: New Practices for Reimagining the City.* Cambridge, MA: MIT Press.

Cuomo, Chris J. 1998. *Feminism and Ecological Communities: An Ethic of Flourishing.* London: Routledge.

Dale, Ole Johan. 2008. "Sustainable City Centre Development: The Singapore City Centre in the Context of Sustainable Development." In *Spatial Planning for a Sustainable Singapore*, edited by Tai-Chee Wong, Belinda Yuen, and Charles Goldblum, 31–57. Dordrecht: Springer Netherlands.

Dali, Faradiah Md, Hafizan Juahir, Ahmad Zaharin Aris, Mohd Ekhwan Toriman, Ai Phing Lim, Ley Juen Looi, Sagoff Syed, and Faradiella Mohd Kusin. 2014. "Assessment of Water Quality Using Environmetric Techniques at Johor River." In *From Sources to Solution*, edited by A. Z. Aris, Tengku Ismail, R. Harun, A. M. Abdullah, and M. Y. Ishak, 403–407. Singapore: Springer.

Daly, Natasha. 2020. "Fake Animal News Abounds on Social Media as Coronavirus Upends Life." *National Geographic*, March 20. https://www.nationalgeographic.com /animals/2020/03/coronavirus-pandemic-fake-animal-viral-social-media-posts/.

Davis, Lucy. 2011. "Zones of Contagion: The Singapore Body Politic and the Body of the Street-Cat." In *Considering Animals*, edited by Carol Freeman, Elizabeth Leane, and Yvette Watt, 183–198. Farnham: Ashgate.

Davison, Aidan. 2008. "The Trouble with Nature: Ambivalence in the Lives of Urban Australian Environmentalists." *Geoforum* 39 (3): 1284–1295. https://doi.org/10.1016 /j.geoforum.2007.06.011.

Davison, Aidan, and Ben Ridder. 2006. "Turbulent Times for Urban Nature: Conserving and Re-Inventing Nature in Australian Cities." *Australian Zoologist* 33 (3): 306–314. https://doi.org/10.7882/az.2006.004.

De Koninck, Rodolphe, Julie Drolet, and Marc Girard. 2008. *Singapore: An Atlas of Perpetual Territorial Transformation*. Singapore: National University of Singapore Press.

Despommier, Dickson D. 2010. *The Vertical Farm: Feeding the World in the 21st Century*. New York: St. Martin's Press.

Dobbs, Stephen. 1994. "'Tongkang, Twakow,' and Lightermen: A People's History of the Singapore River." *Sojourn: Journal of Social Issues in Southeast Asia* 9 (2): 269–276.

Dobbs, Stephen. 2003. *The Singapore River: A Social History, 1819–2002*. Singapore: NUS Press.

Douglas, Ian, and Jonathan P. Sadler. 2011. "Urban Wildlife Corridors: Conduits for Movement or Linear Habitat?" In *The Routledge Handbook of Urban Ecology*, edited by David Goode, Michael C Houck, and Rusong Wang, 274–288. Abingdon: Routledge.

Dromgold, Jacinda R., Caragh G. Threlfall, Briony A. Norton, and Nicholas S. G. Williams. 2020. "Green Roof and Ground-Level Invertebrate Communities Are Similar and Are Driven by Building Height and Landscape Context." *Journal of Urban Ecology* 6 (1): 1–9.

Economist Intelligence Unit. 2011. *Asian Green City Index: Assessing the Environmental Performance of Asia's Major Cities*. Munich: Siemens AG.

Economist Intelligence Unit. 2019. *Global Food Security Index 2019*. London: The Economist Intelligence Unit Limited.

Eke, Joyner, Ahmed Yusuf, Adewale Giwa, and Ahmed Sodiq. 2020. "The Global Status of Desalination: An Assessment of Current Desalination Technologies, Plants and Capacity." *Desalination* 495:114633.

Elangovan, Navene. 2019. "The Big Read: To Build a Strong Water-Saving Culture, S'pore Needs More than Recycled Messages." *Today Online*, March 16. https://www .todayonline.com/big-read/big-read-build-strong-water-saving-culture-singapore -needs-more-recycled-messages.

Elsaid, Khaled, Mohammed Kamil, Enas Taha Sayed, Mohammad Ali Abdelkareem, Tabbi Wilberforce, and A. Olabi. 2020. "Environmental Impact of Desalination Technologies: A Review." *Science of the Total Environment* 748:141528. https://doi.org /10.1016/j.scitotenv.2020.141528.

Emont, Jon. 2020. "Singapore, a City of Skyscrapers and Little Land, Turns to Farming." *The Wall Street Journal*, November 22.

Energy Market Authority. 2020. "Singapore's Energy Story." Energy Market Authority. Singapore.

Engler, Nicholas, and Moncef Krarti. 2021. "Review of Energy Efficiency in Controlled Environment Agriculture." *Renewable and Sustainable Energy Reviews* 141:110786. https://doi.org/10.1016/j.rser.2021.110786.

Environmental Resources Management (S) Pte Ltd. 2016. *Environmental Impact Assessment on Central Catchment Nature Reserve for the Proposed Cross Island Line, Site Investigation Environmental Impact—Assessment Report—Volume III. Alignment Option 1.* Singapore: Land Transport Authority.

Environmental Resources Management (S) Pte Ltd. 2019. *Environmental Impact Assessment on Central Catchment Nature Reserve for the Proposed Cross Island Line, Final Construction & Operation Environmental Impact Assessment Report.* Singapore: Land Transport Authority.

Er, Kenneth, Michelle Lim, and Andrew Grant. 2010. "Gardens by the Bay— Designing a Nation's Garden in the Heart of Singapore's Downtown." *CITYGREEN* 1. Centre for Urban Greenery and Ecology: 32–39.

Er, Kenneth, Ng Boon Gee, Andy Kwek, and Matthew Potter. 2013. "World Building of the Year: Cooled Conservatories at Gardens by the Bay." *CITYGREEN* 6. Centre for Urban Greenery and Ecology: 18–21.

Ernstson, Henrik, and Sverker Sorlin, eds. 2019. *Grounding Urban Natures: Histories and Futures of Urban Ecologies.* Cambridge, MA: MIT Press.

Fabian, Johannes. 1983. *Time and the Other: How Anthropology Makes Its Object.* New York: Columbia University Press.

Fei, Chung Yi, and Chenny Li. 2014. "Eco-Link@BKE—Reconnecting Our Biodiversity." *NParks Buzz* 1 (20). National Parks Board.

Fei, Chung Yi, Wong Tuan Wah, and Sharon Chan. 2016. "Eco-Link@BKE: A Safe Corridor for Our Biodiversity." *CITYGREEN* 12. Centre for Urban Greenery and Ecology: 92–95.

Fisher, Berenice, and Joan Tronto. 1990. "Toward a Feminist Theory of Caring." In *Circles of Care*, edited by Emily K. Abel, and Margaret K. Nelson, 35–62. Albany: State University of New York Press.

Foucault, Michel. 1986. "Of Other Spaces." Translated by Jay Miskowiec. *Diacritics* 16 (1). 22–27.

Franklin, Adrian. 2017. "The More-Than-Human City." *The Sociological Review* 65 (2): 202–217. https://doi.org/10.1111/1467-954x.12396.

Franklin, Sarah. 2001. "Culturing Biology: Cell Lines for the Second Millennium." *Health* 5 (3): 335–354. https://doi.org/10.1177/136345930100500304.

Frazier, Ian. 2017. "The Vertical Farm." *The New Yorker*, January 1. https://www.newyorker.com/magazine/2017/01/09/the-vertical-farm.

Friess, Daniel A. 2017a. "Singapore as a Long-Term Case Study for Tropical Urban Ecosystem Services." *Urban Ecosystems* 20 (2), 277–229. https://doi.org/10.1007/s11252-016-0592-7.

Friess, Daniel A. 2017b. "Mangrove Rehabilitation along Urban Coastlines: A Singapore Case Study." *Regional Studies in Marine Science* 16 (November): 279–289. https://doi.org/10.1016/j.rsma.2017.09.013.

Friess, Daniel A., Daniel R. Richards, and Valerie X. H. Phang. 2016. "Mangrove Forests Store High Densities of Carbon across the Tropical Urban Landscape of Singapore." *Urban Ecosystems* 19 (2): 795–810. https://doi.org/10.1007/s11252-015-0511-3.

Fuentes, Agustín. 2010. "Naturalcultural Encounters in Bali: Monkeys, Temples, Tourists, and Ethnoprimatology." *Cultural Anthropology* 25 (4): 600–624. https://doi.org/10.1111/j.1548-1360.2010.01071.x.

Gaard, Greta. 2001. "Women, Water, Energy: An Ecofeminist Approach." *Organization & Environment* 14 (2): 157–172. https://doi.org/10.1177/1086026601142002.

Gaffen, David. 2020. "When Oil Became Waste: A Week of Turmoil for Crude, and More Pain to Come." *Reuters*, April 26.

Gandy, Matthew. 2003. *Concrete and Clay: Reworking Nature in New York City*. Cambridge, MA: MIT Press.

Gandy, Matthew. 2006. "Urban Nature and the Ecological Imaginary." In *In the Nature of Cities: Urban Political Ecology and the Politics of Urban Metabolism*, edited by Nik Heynen, Maria Kaika, and Erik Swyngedouw. London: Routledge.

Gandy, Matthew. 2014. *The Fabric of Space: Water, Modernity, and the Urban Imagination*. Cambridge, MA: MIT Press.

Gandy, Matthew. 2022a. *Natura Urbana: Ecological Constellations in Urban Space*. Cambridge, MA: MIT Press.

Gandy, Matthew. 2022b. "The Zoonotic City: Urban Political Ecology and the Pandemic Imaginary." *International Journal of Urban and Regional Research* 46 (2): 202–219.

Gardens by the Bay. n.d. "Sustainability Efforts." *Gardens by the Bay.*

Garmendia, Eneko, Evangelia Apostolopoulou, William M. Adams, and Dimitrios Bormpoudakis. 2016. "Biodiversity and Green Infrastructure in Europe: Boundary object or ecological trap?" *Land Use Policy* 56: 315–319. https://doi.org/10.1016/j.landusepol.2016.04.003.

Gattinoni, Paola, Enrico Maria Pizzarotti, and Laura Scesi. 2014. *Engineering Geology for Underground Works.* Dordrecht: Springer. https://doi.org/10.1007/978-94-007-7850-4.

Gaw, Leon Yan-Feng, and Daniel Rex Richards. 2021. "Development of Spontaneous Vegetation on Reclaimed Land in Singapore Measured by NDVI." *PLOS ONE* 16 (1): e0245220. https://doi.org/10.1371/journal.pone.0245220.

Geddie, John, and Edgar Su. 2019. "From Sky Farms to Lab-Grown Shrimp, Singapore Eyes Food Future." *Reuters*, May 30. https://www.reuters.com/article/uk-singapore-agriculture-idUKKCN1T00F0.

George, Cherian. 2000. *Singapore: The Air-Conditioned Nation: Essays on the Politics of Comfort and Control 1990–2000.* Singapore: Landmark Books.

Ginn, Franklin. 2014. "Sticky Lives: Slugs, Detachment and More-Than-Human Ethics in the Garden." *Transactions of the Institute of British Geographers* 39 (4): 532–544. https://doi.org/10.1111/tran.12043.

Ginn, Franklin. 2017. *Domestic Wild: Memory, Nature and Gardening in Suburbia.* Abingdon: Routledge.

Ginn, Franklin, Uli Beisel, and Maan Barua. 2014. "Flourishing with Awkward Creatures: Togetherness, Vulnerability, Killing." *Environmental Humanities* 4 (1): 113–123.

Giraud, Eva, Greg Hollin, Tracey Potts, and Isla Forsyth. 2018. "A Feminist Menagerie." *Feminist Review* 118 (1): 61–79. https://doi.org/10.1057/s41305-018-0103-1.

Goh, Robbie B. H. 2005. *Contours of Culture: Space and Social Difference in Singapore.* Hong Kong: Hong Kong University Press.

Goh, Xue, Mohanasundar Radhakrishnan, Chris Zevenbergen, and Assela Pathirana. 2017. "Effectiveness of Runoff Control Legislation and Active, Beautiful, Clean (ABC) Waters Design Features in Singapore." *Water* 9 (8): 627. https://doi.org/10.3390/w9080627.

Goldfarb, Ben. 2020. "When Wildlife Safety Turns into Fierce Political Debate." *High Country News*, January 1.

Goldfarb, Ben. 2021. "Animals Need Infrastructure, Too." *Vox*, November 12.

Goodman, Wylie, and Jennifer Minner. 2019. "Will the Urban Agricultural Revolution Be Vertical and Soilless? A Case Study of Controlled Environment Agriculture in

New York City." *Land Use Policy* 83: 160–173. https://doi.org/10.1016/j.landusepol
.2018.12.038.

Goodyear-Kaʻōpua, Noelani. 2017. "Protectors of the Future, Not Protestors of the
Past: Indigenous Pacific Activism and Mauna a Wākea." *South Atlantic Quarterly* 116
(1): 184–194. https://doi.org/10.1215/00382876-3749603.

Goossen, Willem Jan. 2018. "Interview (with Williem Jan Goossen)—The Dutch
Make Room for the River." *Europe Environment Agency*. https://www.eea.europa.eu
/signals/signals-2018-content-list/articles/interview-2014-the-dutch-make.

Gottlieb, Robert. 2022. *Care-Centered Politics: From the Home to the Planet*. Cambridge,
MA: MIT Press.

Graham, Stephen. 2015. "Life Support: The Political Ecology of Urban Air." *City* 19
(2–3): 192–215. https://doi.org/10.1080/13604813.2015.1014710.

Graham, Stephen. 2016. *Vertical: The City from Satellites to Bunkers*. New York: Verso.

Graham, Stephen, Renu Desai, and Colin McFarlane. 2013. "Water Wars in Mumbai."
Public Culture 25 (1): 115–141. https://doi.org/10.1215/08992363-1890486.

Granjou, Céline, and Juan Francisco Salazar. 2016. "Future." *Environmental Humanities* 8 (2): 240–244. https://doi.org/10.1215/22011919-3664342.

Greear, Jake P. 2016. "Decentralized Production and Affective Economies: Theorizing the Ecological Implications of Localism." *Environmental Humanities* 7 (1):
107–127. https://doi.org/10.1215/22011919-3616353.

Green, Kelsey, and Franklin Ginn. 2014. "The Smell of Selfless Love: Sharing Vulnerability with Bees in Alternative Apiculture." *Environmental Humanities* 4 (1): 149–170.
https://doi.org/10.1215/22011919-3614971.

Green, Lesley. 2018. "Making It through the Water Crisis." *University of Cape Town*,
January 31.

Green, Lesley. 2020. *Rock/Water/Life: Ecology and Humanities for a Decolonial South
Africa*. Durham, NC: Duke University Press.

Greenfield, Patrick. 2021. "Animal Crossings: The Ecoducts Helping Wildlife
Navigate Busy Roads across the World." *The Guardian*. December 29. https://www
.theguardian.com/environment/2021/dec/29/wildlife-bridges-saving-creatures-big
-and-small-aoe.

Greenhouse, Carol J. 1996. *A Moment's Notice: Time Politics across Cultures*. New York:
Cornell University Press.

Griffiths, Steven, and Benjamin K. Sovacool. 2020. "Rethinking the Future Low-
Carbon City: Carbon Neutrality, Green Design, and Sustainability Tensions in the
Making of Masdar City." *Energy Research & Social Science* 62 (April): 101368. https://
doi.org/10.1016/j.erss.2019.101368.

Griffiths, Tom. 2007. "The Humanities and an Environmentally Sustainable Australia." *Australian Humanities Review* 43 (March).

Gulsrud, Natalie Marie, and Ooi Can-Seng. 2015. "Manufacturing Green Consensus: Urban Greenspace Governance in Singapore." In *Urban Forests, Trees, and Greenspace: A Political Ecology Perspective*, edited by Sandberg L. Sanders, Adrina Bardekjian, and Sadia Butt, 77–92. New York: Routledge.

Günel, Gökçe. 2019. *Spaceship in the Desert: Energy, Climate Change, and Urban Design in Abu Dhabi.* Durham, NC: Duke University Press.

Gunther, Marc. 2018. "Can Deepwater Aquaculture Avoid the Pitfalls of Coastal Fish Farms?" *Yale Environment 360,* January 25.

Haddad, Nick M., Lars A. Brudvig, Ellen I. Damschen, Daniel M. Evans, Brenda L. Johnson, Douglas J. Levey, John L. Orrock et al. 2014. "Potential Negative Ecological Effects of Corridors." *Conservation Biology* 28 (5): 1178–1187.

Hajer, Maarten. 1992. "The Politics of Environmental Performance Review: Choices in Design." In *Achieving Environmental Goals: The Concept and Practice of Environmental Performance Review*, edited by Erik Lykke. London: Belhaven Press.

Hajer, Maarten A. 1997. *The Politics of Environmental Discourse: Ecological Modernization and the Policy Process.* Oxford Academic. https://doi.org/10.1093/019829333x.001.0001.

Hamilton, Andrew J., Kristal Burry, Hoi-Fei Mok, S. Fiona Barker, James R. Grove, and Virginia G. Williamson. 2014. "Give Peas a Chance? Urban Agriculture in Developing Countries. A Review." *Agronomy for Sustainable Development* 34 (1): 45–73. https://doi.org/10.1007/s13593-013-0155-8.

Hamnett, Stephen, and Belinda Yuen. 2019. "Planning Singapore: Challenges and Choices." In *Planning Singapore: The Experimental City*, edited by Stephen Hamnett and Belinda Yuen, 1–36. Abingdon: Routledge.

Han, Kirsten. 2015. "Land-Starved Singapore Exhumes Its Cemeteries to Build Roads and Malls." *The Guardian*, August 7. https://www.theguardian.com/cities/2015/aug/07/land-starved-singapore-exhumes-its-cemeteries-to-build-roads-and-malls.

Hansard. 1978. *Acquisition of Private Cemeteries.* Parliament of Singapore. 7 April. Session 1, vol. 37, sitting 17.

Hansard. 2014. *Singapore's Water Sufficiency During Dry Spell.* Parliament of Singapore. 7 March. Session 1, vol. 91, sitting 10.

Hansard. 2015. *Reclamation of Lim Chu Kang Farm Land for Military Use.* Parliament of Singapore. 19 January. Session 2, vol. 93, sitting 1.

Hansard. 2016. *Factors for Considering Underground Alignments for Cross Island MRT Line.* Parliament of Singapore. 29 February. Session 1, vol. 94, sitting 7.

Hansard. 2017. *Budget: Committee of Supply—Head T (Ministry of National Development).* Parliament of Singapore. 7 March. Session 1, vol. 94, sitting 40.

Haraway, Donna. 1985. "A Manifesto for Cyborgs: Science, Technology, and Socialist Feminism in the 1980s." *Socialist Review*, 80: 65–108.

Haraway, Donna. 1988. "Situated Knowledges: The Science Question in Feminism and the Privilege of Partial Perspective." *Feminist Studies* 14 (3): 575–599.

Haraway, Donna. 1994. "A Game of Cat's Cradle: Science Studies, Feminist Theory, Cultural Studies." *Configurations* 2 (1): 59–71. https://doi.org/10.1353/con.1994.0009.

Haraway, Donna. 2004. *The Haraway Reader*. New York: Routledge.

Haraway, Donna. 2008. *When Species Meet*. Minneapolis: University of Minnesota Press.

Hardy, A., J. Fuller, M. Huijser, A. Kociolek, and M. Evans. 2007. "Evaluation of Wildlife Crossing Structures and Fencing on US Highway 93 Evaro to Polson and Finalization of Evaluation Plan." *Montana Department of Transportation and U. S. Department of Transportation Federal Highway Administration*. Bozeman, MT: Western Transportation Institute.

Hartigan, John. 2015. "Plant Publics: Multispecies Relating in Spanish Botanical Gardens." *Anthropological Quarterly* 88 (2): 481–507. https://doi.org/10.1353/anq.2015.0024.

Hartigan, John. 2017. *Care of the Species: Races of Corn and the Science of Plant Biodiversity*. Minneapolis: University of Minnesota Press.

Harvey, David. 1993. "The Nature of Environment: The Dialectics of Social and Environmental Change." *The Socialist Register* 29, 1–51.

Head, Lesley, Jennifer Atchison, Catherine Phillips, and Kathleen Buckingham. 2014. "Vegetal Politics: Belonging, Practices and Places." *Social & Cultural Geography* 15 (8): 861–870. https://doi.org/10.1080/14649365.2014.973900.

Hee, Limin, and Giok Ling Ooi. 2003. "The Politics of Public Space Planning in Singapore." *Planning Perspectives* 18 (1): 79–103.

Heise, Ursula K. 2008. *Sense of Place and Sense of Planet: The Environmental Imagination of the Global*. Oxford: Oxford University Press.

Henderson, Joan. 2011. "Understanding and Using Built Heritage: Singapore's National Monuments and Conservation Areas." *International Journal of Heritage Studies* 17 (1): 46–61. https://doi.org/10.1080/13527258.2011.524006.

Heng, Geraldine, and Janadas Devan. 1995. "State Fatherhood: The Politics of Nationalism, Sexuality, and Race in Singapore." In *Bewitching Women, Pious Men: Gender and Body Politics in Southeast Asia*, edited by Aihwa Ong and Michael G. Peletz, 195–215. Berkeley: University of California Press.

Heng, Janice, and Grace Chua. 2014. "'Modern Kampung' to Launch in July BTO." *The Straits Times*, April 27. https://str.sg/3dhp.

Heng, Yee-Kuang. 2013. "A Global City in an Age of Global Risks: Singapore's Evolving Discourse on Vulnerability." *Contemporary Southeast Asia* 35 (3): 423–446. https://doi.org/10.1355/cs35-3e.

Herring, Horace. 2006. "Energy Efficiency—A Critical View." *Energy* 31 (1): 10–20. https://doi.org/10.1016/j.energy.2004.04.055.

Hicks, Robin. 2017. "Planet Earth II Ignores Threats to Singapore's Last Forests." *Eco-Business*, March 15. https://www.eco-business.com/opinion/planet-earth-ii-ignores -threats-to-singapores-last-forests/.

Hilty, Jodi A., William Z. Lidicker Jr., and Adina M. Merenlender. 2006. *Corridor Ecology: The Science and Practice of Linking Landscapes for Biodiversity Conservation.* Washington, DC: Island Press.

Hinchliffe, Steve, Matthew B. Kearnes, Monica Degen, and Sarah Whatmore. 2005. "Urban Wild Things: A Cosmopolitical Experiment." *Environment and Planning D: Society and Space* 23 (5): 643–658. https://doi.org/10.1068/d351t.

Hinchliffe, Steve, and Stephanie Lavau. 2013. "Differentiated Circuits: The Ecologies of Knowing and Securing Life." *Environment and Planning D: Society and Space* 31 (2): 259–274. https://doi.org/10.1068/d6611.

Hinchliffe, Steve, and Sarah Whatmore. 2006. "Living Cities: Towards a Politics of Conviviality." *Science as Culture* 15 (2): 123–138. https://doi.org/10.1080 /09505430600707988.

Hoang, Nguyen Tien, and Keiichiro Kanemoto. 2021. "Mapping the Deforestation Footprint of Nations Reveals Growing Threat to Tropical Forests." *Nature Ecology & Evolution* 5 (6): 845–853. https://doi.org/10.1038/s41559-021-01417-z.

Hodgetts, Timothy, and Jamie Lorimer. 2018. "Animals' Mobilities." *Progress in Human Geography* 44 (1): 4–26. https://doi.org/10.1177/0309132518817829.

Höhler, Sabine. 2010. "The Environment as a Life Support System: The Case of Biosphere 2." *History and Technology* 26 (1): 39–58. https://doi.org/10.1080 /07341510903313048.

Höhler, Sabine. 2015. *Spaceship Earth in the Environmental Age 1960–1990.* London: Routledge.

Horst, Megan, Nathan McClintock, and Lesli Hoey. 2017. "The Intersection of Planning, Urban Agriculture, and Food Justice: A Review of the Literature." *Journal of the American Planning Association* 83 (3): 277–295. https://doi.org/10.1080/01944363 .2017.1322914.

Housing & Development Board. 2016. "Unveiling the Masterplan for Tengah: At Home with Nature." *Housing & Development Board*, September 8.

Housing & Development Board. "Public Housing—A Singapore Icon." *Housing & Development Board*. Last modified July 7, 2023. https://www.hdb.gov.sg/about-us /our-role/public-housing-a-singapore-icon.

Houston, Donna. 2019. "Planning in the Shadow of Extinction: Carnaby's Black Cockatoos and Urban Development in Perth, Australia." *Contemporary Social Science*: 1–14. https://doi.org/10.1080/21582041.2019.1660909.

Houston, Donna, Jean Hillier, Diana MacCallum, Wendy Steele, and Jason Byrne. 2018. "Make Kin, Not Cities! Multispecies Entanglements and 'Becoming-World' in Planning Theory." *Planning Theory* 17 (2): 190–212. https://doi.org/10.1177/1473095216688042.

Houston, Donna, Diana McCallum, Wendy Steele, and Jason Byrne. 2016. "Climate Cosmopolitics and the Possibilities for Urban Planning." *Nature and Culture* 11 (3): 259–277. https://doi.org/10.3167/nc.2016.110303.

Howard, Ebenezer. 1965. *Garden Cities of To-Morrow.* (Original work published in 1902). Cambridge, MA: MIT Press.

Howe, Cymene, Jessica Lockrem, Hannah Appel, Edward Hackett, Dominic Boyer, Randal Hall, Matthew Schneider-Mayerson et al. 2016. "Paradoxical Infrastructures: Ruins, Retrofit, and Risk." *Science, Technology, & Human Values* 41 (3): 547–565. https://doi.org/10.1177/0162243915620017.

Huang, Jianli. 2014. "Resurgent Spirits of Civil Society Activism: Rediscovering the Bukit Brown Cemetery in Singapore." *Journal of the Malaysian Branch of the Royal Asiatic Society* 87 (2): 21–45.

Huijser, Marcel P., John W. Duffield, Anthony P. Clevenger, Robert J. Ament, and Pat T. McGowen. 2009. "Cost–Benefit Analyses of Mitigation Measures Aimed at Reducing Collisions with Large Ungulates in the United States and Canada: A Decision Support Tool." *Ecology and Society* 14 (2). https://doi.org/10.5751/es-03000-140215.

Hultman, Martin. 2013. "The Making of an Environmental Hero: A History of Ecomodern Masculinity, Fuel Cells and Arnold Schwarzenegger." *Environmental Humanities* 2 (1): 79–99. https://doi.org/10.1215/22011919-3610360.

Hustak, Carla, and Natasha Myers. 2012. "Involutionary Momentum: Affective Ecologies and the Sciences of Plant/Insect Encounters." *Differences* 23 (3): 74–118. https://doi.org/10.1215/10407391-1892907.

Hutton, Patrick H. 1993. *History as an Art of Memory.* Burlington, VT: University of Vermont.

IEA. 2019. "Southeast Asia Energy Outlook 2019." *The International Energy Agency.* https://www.iea.org/reports/southeast-asia-energy-outlook-2019.

Ikuko, Nishimoto. 1997. "The 'Civilization' of Time: Japan and the Adoption of the Western Time System." *Time & Society* 6 (2–3): 237–259. https://doi.org/10.1177/0961463x97006002007.

Ingold, Tim. 2010. "Footprints through the Weather-World: Walking, Breathing, Knowing." *Journal of the Royal Anthropological Institute* 16 (1): S121–S139. https://doi.org/10.1111/j.1467-9655.2010.01613.x.

Ingold, Tim, and Jo Lee Vergunst. 2008. "Introduction." In *Ways of Walking: Ethnography and Practice on Foot*, edited by Tim Ingold and Jo Lee Vergunst, 1–19. Hampshire, UK: Ashgate Publishing.

International Monetary Fund. 2020. "Fiscal Monitor: Policies to Support People During the COVID-19 Pandemic." *International Monetary Fund*, April.

IPCC. 2018. *Global Warming of 1.5° C*. Edited by Valérie Masson-Delmotte, Panmao Zhai, Hans-Otto Pörtner, Debra Roberts, Jim Skea, Priyadarshi R. Shukla, Anna Pirani et al. Cambridge: Cambridge University Press. https://doi.org/10.1017 /9781009157940.

Irvine, K. N., Tricia Seow, Ka Wai Leong, and Diana Sze Ing Cheong. 2015. "How High's the Water, Mama? A Reflection on Water Resource Education in Singapore." *HSSE Online* 4 (2): 128–162.

Jacobs, Jane. 1961. *The Death and Life of Great American Cities*. New York: Vintage Books.

Jain, Anuj, Kwek Yan Chong, Marcus Aik Hwee Chua, and Gopalasamy Reuben Clements. 2014. "Moving Away from Paper Corridors in Southeast Asia." *Conservation Biology* 28 (4): 889–891. https://doi.org/10.1111/cobi.12313.

Jalais, Annu. 2021. "The Singapore 'Garden City': The Death and Life of Nature in an Asian City." In *Death and Life of Nature in Asian Cities*, edited by Anne Rademacher and K. Sivaramakrishnan, 1st ed., 82–101. Hong Kong: Hong Kong University Press.

Jeswani, Harish K., Andrew Chilvers, and Adisa Azapagic. 2020. "Environmental Sustainability of Biofuels: A Review." *Proceedings of the Royal Society A: Mathematical, Physical and Engineering Sciences* 476 (2243). https://doi.org/10.1098/rspa.2020.0351.

Jevons, William Stanley. 1865. *The Coal Question; an Inquiry Concerning the Progress of the Nation, and the Probable Exhaustion of Our Coal Mines*. London: Macmillan.

Jiang, Xiaowei, and Ruoqi Wang. 2022. "Wildlife Trade Is Likely the Source of SARS-CoV-2." *Science* 377 (6609): 925–926. https://doi.org/10.1126/science.add8384.

Jones, Edward, Manzoor Qadir, Michelle T. H. van Vliet, Vladimir Smakhtin, and Seong-mu Kang. 2019. "The State of Desalination and Brine Production: A Global Outlook." *Science of the Total Environment* 657: 1343–1356. https://doi.org/10.1016/j .scitotenv.2018.12.076.

Joshi, Yugal Kishore, Cecilia Tortajada, and Asit K. Biswas. 2012. "Cleaning of the Singapore River and Kallang Basin in Singapore: Human and Environmental Dimensions." *Ambio* 41 (7): 777–781. https://doi.org/10.1007/s13280-012-0279-0.

Juanda, Iliyas. 2016. "Tengah Town to Be Built Using Smart, Sustainable Tech." *Today Online*, September 9. https://www.todayonline.com/singapore/hdb-use-smart -technology-keep-tengah-cool-and-green.

Judith, Kate. 2020. "How Mangroves Story: On Being a Filter Feeder." *Swamphen: A Journal of Cultural Ecology*. 7:1–10.

Kaika, Maria. 2004. "Interrogating the Geographies of the Familiar: Domesticating Nature and Constructing the Autonomy of the Modern Home." *International Journal*

of Urban and Regional Research 28 (2): 265–286. https://doi.org/10.1111/j.0309-1317
.2004.00519.x.

Kaika, Maria. 2005. *City of Flows: Modernity, Nature, and the City.* New York: Routledge.

Kantamneni, Neeta. 2020. "The Impact of the COVID-19 Pandemic on Marginalized
Populations in the United States: A Research Agenda." *Journal of Vocational Behavior*
119 (103439). https://doi.org/10.1016/j.jvb.2020.103439.

Kearnes, Matthew, and Lauren Rickards. 2017. "Earthly Graves for Environmental
Futures: Techno-Burial Practices." *Futures* 92 (September): 48–58. https://doi.org/10
.1016/j.futures.2016.12.003.

Keeve, Christian B. 2020. "Fugitive Seeds." *Edge Effects*, February 26. https://
edgeeffects.net/fugitive-seeds/.

Keil, Roger. 2003. "Urban Political Ecology." *Urban Geography* 24 (8): 723–738.

Keil, Roger. 2005. "Progress Report: Urban Political Ecology." *Urban Geography* 26 (7):
640–651.

Keil, Roger. 2020. "An Urban Political Ecology for a World of Cities." *Urban Studies*
57 (11): 2357–2370. https://doi.org/10.1177/0042098020919086.

Khoo, Hong Meng. 2020. *Sky Urban Solutions—Vertical Farming—an Urban Agriculture Solution.* Nanyang Technoprenuership Case Center, Nanyang Technological
University.

Khoo, Teng Chye. 2016. *Urban Solutions: City of Gardens and Water.* Singapore:
Centre for Liveable Cities.

Kirksey, S. Eben, Nicholas Shapiro, and Maria Brodine. 2013. "Hope in Blasted
Landscapes." *Social Science Information* 52 (2): 228–256. https://doi.org/10.1177
/0539018413479468.

Knox, Hannah. 2017. "Affective Infrastructures and the Political Imagination."
Public Culture 29 (2): 363–384. https://doi.org/10.1215/08992363-3749105.

Koelle, Alexandra. 2012. "Intimate Bureaucracies: Roadkill, Policy, and Fieldwork on the
Shoulder." *Hypatia* 27 (3): 651–669. https://doi.org/10.1111/j.1527-2001.2012.01295.x.

Koh, Poh Koon. 2017. "Adopting Technology in Farming Is Not Just for Greater
Productivity; It Is about the Existential Survival of Our Farming Industry." *Facebook Status Update*, September 10. https://www.facebook.com/drkohpohkoon/posts
/adopting-technology-in-farming-is-not-just-for-greater-productivity-it-is-about
-/2054868221205854.

Kohn, Eduardo. 2013. *How Forests Think: Toward an Anthropology beyond the Human.*
Berkeley: University of California Press.

Kong, Lily, and Orlando Woods. 2018. "The Ideological Alignment of Smart Urbanism in Singapore: Critical Reflections on a Political Paradox." *Urban Studies* 55 (4):
679–701. https://doi.org/10.1177/0042098017746528.

Kong, Lily, and Brenda S. A. Yeoh. 1994. "Urban Conservation in Singapore: A Survey of State Policies and Popular Attitudes." *Urban Studies* 31 (2): 247–265. https://doi.org/10.1080/00420989420080231.

Kong, Lily, and Brenda S. A. Yeoh. 2003. *The Politics of Landscapes in Singapore: Constructions of "Nation."* New York: Syracuse University Press.

Kowarik, Ingo. 2020. "Urban Cemeteries in Berlin and Beyond: Life in the Grounds of the Dead." In *The Botanical City*, edited by Matthew Gandy and Sandra Jasper, 305–311. Berlin: Jovis Verlag GmbH.

Krauss, Ken W., Karen L. McKee, Catherine E. Lovelock, Donald R. Cahoon, Neil Saintilan, Ruth Reef, and Luzhen Chen. 2014. "How Mangrove Forests Adjust to Rising Sea Level." *New Phytologist* 202 (1): 19–34.

Kroll, Gary. 2015. "An Environmental History of Roadkill: Road Ecology and the Making of the Permeable Highway." *Environmental History* 20 (1): 4–28. https://doi.org/10.1093/envhis/emu129.

Krugman, Paul. 2020. "Australia Shows Us the Road to Hell." *The New York Times*, January 9. https://www.nytimes.com/2020/01/09/opinion/australia-fires.html.

Kwang, Han Fook, Warren Fernandez, and Sumiko Tan. 2015. *Lee Kuan Yew: The Man and His Ideas.* Singapore: Marshall Cavendish Editions.

Lai, Chee Kien. 2018. "Architect, Historian, Lai Chee Kien." Interviewed by Bharati Jagdis (podcast). *Channel NewsAsia*.

Lai, Samantha, Lynette H. L. Loke, Michael J. Hilton, Tjeerd J. Bouma, and Peter A. Todd. 2015. "The Effects of Urbanisation on Coastal Habitats and the Potential for Ecological Engineering: A Singapore Case Study." *Ocean & Coastal Management* 103 (January): 78–85. https://doi.org/10.1016/j.ocecoaman.2014.11.006.

Land Transport Authority. 2013. *Land Transport Master Plan 2013.* Singapore: Land Transport Authority.

Land Transport Authority. 2016. *Mitigating Measures to Protect the CCNR during SI Works.* Singapore: Land Transport Authority.

Land Transport Authority. 2019. *Land Transport Master Plan 2040.* Singapore: Land Transport Authority.

Lanzoni, Will, and Kyle Almond. 2020. "With Cities on Lockdown, Animals Are Finding More Room to Roam." *CNN*, May 1. https://www.cnn.com/2020/05/01/world/gallery/animals-coronavirus-trnd.

Larkin, Brian. 2013. "The Politics and Poetics of Infrastructure." *Annual Review of Anthropology* 42: 327–343.

Latour, Bruno. 2004. *Politics of Nature: How to Bring the Sciences into Democracy.* Cambridge, MA: Harvard University Press.

Laurance, William F., Gopalasamy Reuben Clements, Sean Sloan, Christine S. O'Connell, Nathan D. Mueller, Miriam Goosem, Oscar Venter et al. 2014. "A Global Strategy for Road Building." *Nature* 513 (7517): 229–232.

Lavis, Anna, Emma-Jayne Abbots, and Luci Attala. 2016. "Reflecting on the Embodied Intersections of Eating and Caring." In *Careful Eating: Bodies, Food and Care*, edited by Emma-Jayne Abbots, Anna Lavis, and Luci Attala, 1–21. London: Routledge.

Lea, Tess. 2020. *Darwin*. Sydney, New South Wales: NewSouth Publishing.

Lee, Hannah, and Thai Pin Tan. 2016. "Singapore's Experience with Reclaimed Water: NEWater." *International Journal of Water Resources Development* 32 (4): 611–621. https://doi.org/10.1080/07900627.2015.1120188.

Lee, Hsien Loong. 2012. "Speech by Prime Minister Lee Hsien Loong at Opening of SIWW-WCS-CES." *Prime Minister's Office Singapore*, July 1.

Lee, Hsien Loong. 2013. "Speech by Prime Minister Lee Hsien Loong at the Official Opening of Tuaspring Desalination Plant." *Prime Minister's Office Singapore*, September 18.

Lee, Hsien Loong. 2014. "Speech by Prime Minister Lee Hsien Loong at the Opening of Exxonmobil's Chemical Plant Expansion on 8 January 2014, 11.30am, at Jurong Island." *Prime Minister's Office Singapore*.

Lee, Hsien Loong. 2019. "National Day Rally 2019." August 18. *Prime Minister's Office Singapore*.

Lee, Kuan Yew. 1965. "Speech Made by the Prime Minister Mr. Lee Kuan Yew at the Official Opening of the Underground Car Park and Public Garden at Raffles Place at 11.00 a.m. On November 27, 1965." National Archives of Singapore.

Lee, Kuan Yew. 1995. "Speech by Mr. Lee Kuan Yew, Senior Minister at the Launch of the National Orchid Garden on Friday, 20 October 1995 at 6.00 p.m. at the Singapore Botanic Gardens." National Archives of Singapore.

Lee, Kuan Yew. 2000. *From Third World to First: The Singapore Story: 1965–2000*. New York: HarperCollins.

Lenzen, Manfred, Ya-Yen Sun, Futu Faturay, Yuan-Peng Ting, Arne Geschke, and Arunima Malik. 2018. "The Carbon Footprint of Global Tourism." *Nature Climate Change* 8 (6): 522–528. https://doi.org/10.1038/s41558-018-0141-x.

Leow, Joanne. 2010. "The Future of Nostalgia: Reclaiming Memory in Tan Pin Pin's Invisible City and Alfian Sa'at's a History of Amnesia." *The Journal of Commonwealth Literature* 45 (1): 115–130. https://doi.org/10.1177/0021989409359859.

Leow, Joanne. 2011. "Placenames and Poetry: Urban Memory in Contemporary Singapore." *Quarterly Literary Review Singapore* 10 (3). http://www.qlrs.com/essay.asp?id=848.

Leow, Joanne. 2012. "On Supertrees, Neo-Colonialism and Globalisation." *Yawning Bread*, July 5.

Lewis III, Roy R. 2005. "Ecological Engineering for Successful Management and Restoration of Mangrove Forests." *Ecological Engineering* 24 (4): 403–418.

Liao, Kuei-Hsien. 2019. "The Socio-Ecological Practice of Building Blue-Green Infrastructure in High-Density Cities: What Does the ABC Waters Program in Singapore Tell Us?" *Socio-Ecological Practice Research* 1 (1): 67–81. https://doi.org/10.1007/s42532-019-00009-3.

Lidskog, Rolf, and Ingemar Elander. 2012. "Ecological Modernization in Practice? The Case of Sustainable Development in Sweden." *Journal of Environmental Policy & Planning* 14 (4): 411–427. https://doi.org/10.1080/1523908x.2012.737234.

Liew, Kai Khiun, and Natalie Pang. 2015. "Fuming and Fogging Memories: Civil Society and State in Communication of Heritage in Singapore in the Cases of the Singapore Memories Project and the 'Marxist Conspiracy' of 1987." *Continuum* 29 (4): 549–560. https://doi.org/10.1080/10304312.2015.1051806.

Lim, Allan and Peter Barber. 2018. "Rooftop Farms around Singapore: Comcrop's Vision to Harvest New Ideas." Interviewed by Qiao Wei Tay. *Periscope: A JTC Magazine*, December 13.

Lim, Eng-Beng. 2014. "Future Island." *Third Text* 28 (4–5): 443–453. https://doi.org/10.1080/09528822.2014.929879.

Lim, Han She, and Xi Xi Lu. 2016. "Sustainable Urban Stormwater Management in the Tropics: An Evaluation of Singapore's ABC Waters Program." *Journal of Hydrology* 538 (July): 842–862. https://doi.org/10.1016/j.jhydrol.2016.04.063.

Lim, Jessica. 2012. "Gardens by the Bay Opens on June 29; Cooled Conservatories Highlights of Phase One." *The Straits Times*. April 4.

Lim, Tin Seng. 2013. "Singapore: A City of Campaigns." *BiblioAsia* 8 (4) (January–March): 4–7.

Lin, Weiqiang. 2012. "Wasting Time? The Differentiation of Travel Time in Urban Transport." *Environment and Planning A: Economy and Space* 44 (10): 2477–2492. https://doi.org/10.1068/a4525.

Lindner, Christoph, and Miriam Meissner. 2019. "Urban Imaginaries in Theory and Practice." In *The Routledge Companion to Urban Imaginaries*, edited by Christoph Lindner and Miriam Meissner, 1–22. New York: Routledge.

Linton, Jamie. 2010. *What Is Water? The History of a Modern Abstraction*. Vancouver: UBC Press.

Linton, Jamie, and Jessica Budds. 2014. "The Hydrosocial Cycle: Defining and Mobilizing a Relational-Dialectical Approach to Water." *Geoforum* 57 (57): 170–180. https://doi.org/10.1016/j.geoforum.2013.10.008.

Loftus, Alex. 2009. "Rethinking Political Ecologies of Water." *Third World Quarterly* 30 (5): 953–968. https://doi.org/10.1080/01436590902959198.

Loh, Kah Seng. 2009. "Conflict and Change at the Margins: Emergency Kampong Clearance and the Making of Modern Singapore." *Asian Studies Review* 33 (2): 139–159. https://doi.org/10.1080/10357820902923258.

Loke, Lynette H. L., Tjeerd J. Bouma, and Peter A. Todd. 2017. "The Effects of Manipulating Microhabitat Size and Variability on Tropical Seawall Biodiversity: Field and Flume Experiments." *Journal of Experimental Marine Biology and Ecology* 492 (July): 113–120. https://doi.org/10.1016/j.jembe.2017.01.024.

Lorimer, Jamie. 2007. "Nonhuman Charisma." *Environment and Planning D: Society and Space* 25 (5): 911–932. https://doi.org/10.1068/d71j.

Low, Kelvin E. Y. 2017. "Concrete Memories and Sensory Pasts: Everyday Heritage and the Politics of Nationhood." *Pacific Affairs* 90 (2): 275–295. https://doi.org/10.5509/2017902275.

Low, Tim. 2013. "Considering Corridors." *Wildlife Australia* 50 (4): 4–8.

Lowenthal, David. 1998. *The Heritage Crusade and the Spoils of History*. Cambridge: Cambridge University Press.

Lyons-Lee, Lenore. 1998. "The 'Graduate Woman' Phenomenon: Changing Constructions of the Family in Singapore." *Sojourn: Journal of Social Issues in Southeast Asia* 13 (2): 309–27.

Macnaghten, Phil, and John Urry. 1998. *Contested Natures*. London: Sage Publications.

Mam, Kalyanee. 2018. "Lost World." *Emergence Magazine*, July 18. https://emergencemagazine.org/film/lost-world/.

March, Hug. 2015. "The Politics, Geography, and Economics of Desalination: A Critical Review." *Wires Water* 2 (3): 231–243. https://doi.org/10.1002/wat2.1073.

Marder, Michael. 2013. *Plant-Thinking: A Philosophy of Vegetal Life*. New York: Columbia University Press.

Martin, Aryn, Natasha Myers, and Ana Viseu. 2015. "The Politics of Care in Technoscience." *Social Studies of Science* 45 (5): 625–641. https://doi.org/10.1177/0306312715602073.

Marx, Karl. 1973. *Grundrisse: Foundations of the Critique of Political Economy*. Translated by Martin Nicolaus. London: Penguin Books.

Mastercard. 2019. *Global Destination Cities Index 2019*. Mastercard.

McClintock, Nathan. 2010. "Why Farm the City? Theorizing Urban Agriculture through a Lens of Metabolic Rift." *Cambridge Journal of Regions, Economy and Society* 3 (2): 191–207. https://doi.org/10.1093/cjres/rsq005.

McFarlane, Colin. 2011. "Assemblage and Critical Urbanism." *City* 15 (2): 204–224. https://doi.org/10.1080/13604813.2011.568715.

McFarlane, Colin, and Jonathan Rutherford. 2008. "Political Infrastructures: Governing and Experiencing the Fabric of the City." *International Journal of Urban and Regional Research* 32 (2): 363–374. https://doi.org/10.1111/j.1468-2427.2008.00792.x.

McGrath, Matt. 2020. "Coronavirus: Air Pollution and CO2 Fall Rapidly as Virus Spreads." BBC, March 19. https://www.bbc.com/news/science-environment-51944780.

McNeill, Donald. 2019. "Volumetric Urbanism: The Production and Extraction of Singaporean Territory." *Environment and Planning A: Economy and Space* 51 (4): 849–868.

Medley, Graham. 2020. Interview with BBC Newsnight. "Coronavirus: Can Herd Immunity Protect the Population?" BBC Newsnight, March 13.

Meerganz von Medeazza, Gregor. 2005. "'Direct' and Socially-Induced Environmental Impacts of Desalination." *Desalination* 185 (1–3): 57–70. https://doi.org/10.1016/j.desal.2005.03.071.

Melo Zurita, Maria de Lourdes, Paul George Munro, and Donna Houston. 2018. "Un-Earthing the Subterranean Anthropocene." *Area* 50 (3): 298–305. https://doi.org/10.1111/area.12369.

Ministry of Law. 2015. "Legislative Changes to Facilitate Future Planning and Development of Underground Space." Press Releases. February 12. https://mlaw.gov.sg/news/press-releases/legislative-changes-planning-development-underground-space/.

Ministry of the Environment and Water Resources and Ministry of National Development. 2014. *Sustainable Singapore Blueprint 2015*. Singapore: Ministry of the Environment and Water Resources, Ministry of National Development.

Ministry of the Environment and Water Resources and Ministry of National Development. 2016a. *Take Action Today, for a Sustainable Future. Singapore*. Ministry of the Environment and Water Resources, Ministry of National Development.

Ministry of the Environment and Water Resources and Ministry of National Development. 2016b. *A Climate-Resilient Singapore, for a Sustainable Future*. Singapore: Ministry of the Environment and Water Resources, Ministry of National Development.

Ministry of National Development. 2018. "Singapore and China to Jointly Promote the Replication of the Tianjin Eco-City's Experience." *Singapore Ministry of National Development*, September 20.

Ministry of Trade and Industry. 1984. "Report of the Tourism Task Force." *Ministry of Trade and Industry*.

Ministry of Transport. 2019. "Cross Island Line to Run 70 Metres under Central Catchment Nature Reserve." *Ministry of Transport*, December 4. https://www.mot.gov.sg/news/Details/cross-island-line-to-run-70-metres-under-central-catchment-nature-reserve/.

Mok, Hoi-Fei, Virginia G. Williamson, James R. Grove, Kristal Burry, S. Fiona Barker, and Andrew J. Hamilton. 2014. "Strawberry Fields Forever? Urban Agriculture in Developed Countries: A Review." *Agronomy for Sustainable Development* 34: 21–43. https://doi.org/10.1007/s13593-013-0156-7.

Mol, Annemarie. 2008. *The Logic of Care: Health and the Problem of Patient Choice.* London: Routledge.

Mol, Arthur P. J. 1995. *The Refinement of Production: Ecological Modernization Theory and the Chemical Industry.* Utrecht: International Books/Van Arkel.

Mol, Arthur P. J. 2002. "Ecological Modernization and the Global Economy." *Global Environmental Politics* 2 (2): 92–115. https://doi.org/10.1162/15263800260047844.

Mol, Arthur P. J., and David A. Sonnenfeld. 2000. "Ecological Modernisation around the World: An Introduction." *Environmental Politics* 9 (1): 1–14. https://doi.org/10.1080/09644010008414510.

Mol, Arthur P. J., and Gert Spaargaren. 2000. "Ecological Modernisation Theory in Debate: A Review." *Environmental Politics* 9 (1): 17–49. https://doi.org/10.1080/09644010008414511.

Montesclaros, Jose Ma, Stella Liu, and Paul P. Teng. 2018. "Scaling up Commercial Urban Agriculture to Meet Food Demand in Singapore." *NTS Report* (7). Singapore: NTS Centre.

Moore, Jason W. 2015. *Capitalism in the Web of Life: Ecology and the Accumulation of Capital,* 1st ed. London: Verso.

Moore, Oliver. 2019. "Vertical Farming Shoots . . . Organic in the Foot?" *Agricultural and Rural Convention 2020,* June 21.

Morgan, Kevin, Terry Marsden, and Jonathan Murdoch. 2008. *Worlds of Food: Place, Power, and Provenance in the Food Chain.* Oxford: Oxford University Press.

Mougeot, Luc J.A. 2000. "Urban Agriculture: Definition, Presence, Potentials and Risks, and Policy Challenges." *International Development Research Centre (IDRC)* 62. Ottawa: International Development Research Centre (IDRC).

Muller, Adrian, Marie Ferré, Stefanie Engel, Andreas Gattinger, Annelie Holzkämper, Robert Huber, Moritz Müller, and Johan Six. 2017. "Can Soil-Less Crop Production Be a Sustainable Option for Soil Conservation and Future Agriculture?" *Land Use Policy* 69 (December): 102–105. https://doi.org/10.1016/j.landusepol.2017.09.014.

Murphy, Michelle. 2015. "Unsettling Care: Troubling Transnational Itineraries of Care in Feminist Health Practices." *Social Studies of Science* 45 (5): 717–737. https://doi.org/10.1177/0306312715589136.

Myers, Natasha. 2015. "Edenic Apocalypse: Singapore's End-of-Time Botanical Tourism." In *Art in the Anthropocene: Encounters among Aesthetics, Politics, Environments and Epistemologies,* edited by Heather M. Etienne and Etienne Turpin, 31–42. London: Open Humanities Press.

Myers, Natasha. 2017. "From the Anthropocene to the Planthroposcene: Designing Gardens for Plant/People Involution." *History and Anthropology* 28 (3): 297–301. https://doi.org/10.1080/02757206.2017.1289934.

Myers, Natasha. 2019. "From Edenic Apocalypse to Gardens against Eden: Plants and People in and after the Anthropocene." In *Infrastructure, Environment, and Life in the Anthropocene*, edited by Kregg Hetherington, 115–148. Durham, NC: Duke University Press.

National Environment Agency. 2018. *Singapore's Fourth National Communication and Third Biennial Update Report.* National Environment Agency. December.

National Parks Board. 2010. "Singapore Index formally endorsed as a biodiversity measurement tool for cities." National Parks Board, Singapore, October 30.

National Parks Board. 2011. "Gardens by the Bay." National Parks Board, Singapore, February 17.

National Parks Board. 2012. "A Guide to Sungei Buloh Wetland Reserve." National Parks Board, Singapore. https://www.nparks.gov.sg/-/media/nparks-real-content/learning/learning-journeys/guided-walks/diy-guided-walks/sungei-buloh/diy-trail-guide.pdf.

National Parks Board. 2018. "Central Catchment Nature Reserve." National Parks Board, Singapore.

National Population and Talent Division. 2013. *A Sustainable Population for a Dynamic Singapore: Population White Paper.* Prime Minister's Office.

The Nature of Cities. 2015. "Roundtable: Do Urban Green Corridors 'Work'?" *The Nature of Cities.* https://bit.ly/345VSXU.

Nature Society (Singapore). 2010. *Nature Society's Feedback on the Gardens by the Bay Designs.* Nature Society (Singapore).

Nature Society (Singapore). 2013. *Cross Island Line: Discussion and Position Paper.* Nature Society (Singapore). https://www.nss.org.sg/documents/(NSS)%20Cross-Island%20Line%20Position%20Paper.pdf.

Nature Society (Singapore). 2016. *Nature Society's Position Paper on the Mandai Safari Park Holdings (MSPH)'S Mandai Development Plan.* Nature Society (Singapore).

Nature Society (Singapore). 2018. *Nature Society Singapore (NSS)'s Position on HDB's Tengah Forest Plan.* Nature Society (Singapore).

Nature Society (Singapore). 2019. *Nature Group's Response to the CRL Phase II EIA.* Nature Society (Singapore).

Neimanis, Astrida. 2014. "Alongside the Right to Water, a Posthumanist Feminist Imaginary." *Journal of Human Rights and the Environment* 5 (1): 5–24. https://doi.org/10.4337/jhre.2014.01.01.

Neimanis, Astrida. 2017. *Bodies of Water: Posthuman Feminist Phenomenology*. London: Bloomsbury Academic. https://doi.org/10.5040/9781474275415.

Neo, Harvey. 2007. "Challenging the Developmental State: Nature Conservation in Singapore." *Asia Pacific Viewpoint* 48 (2): 186–199. https://doi.org/10.1111/j.1467-8373.2007.00340.x.

Neo, Harvey. 2010. "Unpacking the Postpolitics of Golf Course Provision in Singapore." *Journal of Sport and Social Issues* 34 (3): 272–287. https://doi.org/10.1177/0193723510377313.

Newman, Peter. 2014. "Biophilic Urbanism: A Case Study on Singapore." *Australian Planner* 51 (1): 47–65. https://doi.org/10.1080/07293682.2013.790832.

Newman, Peter. 2019. "The Challenge of Climate Change for Singapore." In *Planning Singapore: The Experimental City*, edited by Stephen Hamnett and Belinda Yuen, 151–169. Abingdon: Routledge.

Ng, Peter Joo Hee. 2018. "Singapore: Transforming Water Scarcity into a Virtue." In *Assessing Global Water Megatrends. Water Resources Development and Management*, edited by Asit K. Biswas, Cecilia Tortajada, and Philippe Rohner, 179–196. Singapore: Springer.

Ng, Weng Hoong. 2012. *Singapore, the Energy Economy: From the First Refinery to the End of Cheap Oil, 1960–2010*. New York: Routledge.

Nixon, Rob. 2019. "The Anthropocene: Promise and Pitfalls of an Epochal Idea." *Edge Effects*, October 12. https://edgeeffects.net/anthropocene-promise-and-pitfalls/.

O'Dempsey, Tony. 2014. "Singapore's Changing Landscape since C. 1800." In *Nature Contained: Environmental Histories of Singapore*, edited by Timothy P. Barnard, 17–48. Singapore: NUS Press.

Ogden, Lesley Evans. 2015. "Do Wildlife Corridors Have a Downside?" *BioScience* 65 (4): 452. https://doi.org/10.1093/biosci/biv021.

Ogle, Vanessa. 2015. *The Global Transformation of Time: 1870-1950*. Cambridge, MA: Harvard University Press.

O'Gorman, Emily. 2017. "Imagined Ecologies: A More-Than-Human History of Malaria in the Murrumbidgee Irrigation Area, New South Wales, Australia, 1919–45." *Environmental History* 22 (3): 486–514. https://doi.org/10.1093/envhis/emx056.

O'Gorman, Emily. 2021. *Wetlands in a Dry Land: More-than-human Histories of Australia's Murray-Darling Basin*. Seattle: University of Washington Press.

O'Gorman, Emily, and Andrea Gaynor, 2020. "More-Than-Human Histories." *Environmental History* 25:711–735.

Ong, Aihwa. 2016. *Fungible Life: Experiment in the Asian City of Life*. Durham, NC: Duke University Press.

Ong, Corinne, Lyle Fearnley, Quek Ri An, and Chia Siow Boon. 2019. "Recycling Water and Waste in Singapore." In *Planning Singapore: The Experimental City*, edited by Stephen Hamnett and Belinda Yuen, 130–150. Abingdon: Routledge.

Ong, Lauren. 2019. "More Locally Grown Organic Food in Store after Farm Gets First-of-Its-Kind Certification." *Today Online*, June 11. https://www.todayonline.com /singapore/more-locally-grown-organic-food-set-hit-supermarket-shelves-after-farm -here-wins-first.

O'Reilly, Kathleen. 2006. "'Traditional' Women, 'Modern' Water: Linking Gender and Commodification in Rajasthan, India." *Geoforum* 37 (6): 958–972. https://doi .org/10.1016/j.geoforum.2006.05.008.

Orsini, Francesco, Marielle Dubbeling, Henk de Zeeuw, and Giorgio Gianquinto. 2017. *Rooftop Urban Agriculture*. Springer EBooks. Springer International Publishing. https://doi.org/10.1007/978-3-319-57720-3.

Oswin, Natalie. 2010. "The Modern Model Family at Home in Singapore: A Queer Geography." *Transactions of the Institute of British Geographers* 35 (2): 256–268.

Oswin, Natalie. 2019. *Global City Futures: Desire and Development in Singapore*. Athens: University of Georgia Press.

Pak, Hui Ying, C. Joon Chuah, Ee Ling Yong, and Shane A. Snyder. 2021. "Effects of Land Use Configuration, Seasonality and Point Source on Water Quality in a Tropical Watershed: A Case Study of the Johor River Basin." *Science of the Total Environment* 780: 146661. https://doi.org/10.1016/j.scitotenv.2021.146661.

Pak, Hui Ying, C. Joon Chuah, Ee Ling Yong, and Shane A. Snyder., Clare. 2003. "Placing Animals in Urban Environmental Ethics." *Journal of Social Philosophy* 34 (1): 64–78. https://doi.org/10.1111/1467-9833.00165.

Panagopoulos, Argyris, and Katherine-Joanne Haralambous. 2020. "Environmental Impacts of Desalination and Brine Treatment-Challenges and Mitigation Measures." *Marine Pollution Bulletin* 161:111773.

Paschold, Anja, Rayko Halitschke, and Ian T. Baldwin. 2006. "Using 'Mute' Plants to Translate Volatile Signals." *The Plant Journal* 45:275–291. https://doi.org/10.1111/j .1365-313x.2005.02623.x.

Perks, Samuel. 2017. "'Here's to the Grass We Step On!': Complicating the Spatial Dynamics of the Garden City in Singaporean Historical Fiction." *Journal of Postcolonial Writing* 53 (6): 673–685. https://doi.org/10.1080/17449855.2017.1401199.

Perry, Martin, Lily Kong, and Brenda Yeoh. 1997. *Singapore: A Developmental City State*. Chichester: Wiley.

Pilkey, Orrin H., Norma J. Longo, William J. Neal, Nelson G. Rangel-Buitrago, Keith C. Pilkey, and Hannah L. Hayes. 2022. *Vanishing Sands: Losing Beaches to Mining*. Durham, NC: Duke University Press.

Pin, Wan Wee. 2014. "Memory and the Nation: On the Singapore Memory Project." *Alexandria*, *25*(3): 63–70. https://doi.org/10.7227/ALX.0031.

Plumwood, Val. 1993. *Feminism and the Mastery of Nature*. London: Routledge.

Plumwood, Val. 2002. *Environmental Culture: The Ecological Crisis of Reason*. London: Routledge.

Plumwood, Val. 2005. "Decolonising Australian Gardens: Gardening and the Ethics of Place." *Australian Humanities Review* 36 (July): 1–9.

Plumwood, Val. 2008. "Shadow Places and the Politics of Dwelling." *Australian Humanities Review* 44 (March): 139–150.

Plumwood, Val. 2009. "Nature in the Active Voice." *Australian Humanities Review* 46 (May). https://doi.org/10.22459/ahr.46.2009.10.

Plumwood, Val. 2012. "Animals and Ecology: Towards a Better Integration." In *The Eye of the Crocodile*, edited by Lorraine Shannon, 77–90. Canberra: Australian National University Press. https://doi.org/10.22459/ec.11.2012.

Poon, Angelia. 2009. "Pick and Mix for a Global City: Race and Cosmopolitanism in Singapore." In *Race and Multiculturalism in Malaysia and Singapore*, edited by Daniel P. S. Goh, Matilda Gabrielpillai, Philip Holden, and Gaik Cheng Khoo, 84–99. London: Routledge.

Pow, Choon Piew. 2009. "Public Intervention, Private Aspiration: Gated Communities and the Condominisation of Housing Landscapes in Singapore." *Asia Pacific Viewpoint* 50 (2): 215–227. https://doi.org/10.1111/j.1467-8373.2009.01394.x.

Pow, Choon Piew. 2014. "License to Travel: Policy Assemblage and the 'Singapore Model.'" *City* 18 (3): 287–306. https://doi.org/10.1080/13604813.2014.908515.

Pow, Choon Piew, and Harvey Neo. 2015. "Modelling Green Urbanism in China." *Area* 47 (2): 132–140. https://doi.org/10.1111/area.12128.

Powell, Miles Alexander. 2019. "Singapore's Lost Coast: Land Reclamation, National Development and the Erasure of Human and Ecological Communities, 1822–Present." *Environment and History* 27 (4): 635–663. https://doi.org/10.3197/096734019x15631846928710.

Poynter, Jane. 2006. *The Human Experiment: Two Years and Twenty Minutes inside Biosphere 2*. New York: Basic Books.

Probyn, Elspeth. 2014. "Women Following Fish in a More-Than-Human World." *Gender, Place & Culture*, 21 (5): 589–603.

Public Utilities Board. 2018a. *Active, Beautiful, Clean Waters Design Guidelines 4th Edition*. *Public Utilities Board*.

Public Utilities Board. 2018b. "Our Water Our Future." *Public Utilities Board*.

Public Utilities Board. 2023. "Marina Barrage." *Public Utilities Board*.

Puig de la Bellacasa, Maria. 2011. "Matters of Care in Technoscience: Assembling Neglected Things." *Social Studies of Science* 41 (1): 85–106. https://doi.org/10.1177 /0306312710380301.

Puig de la Bellacasa, Maria. 2012. "'Nothing Comes without Its World': Thinking with Care." *The Sociological Review* 60 (2): 197–216. https://doi.org/10.1111/j.1467 -954x.2012.02070.x.

Puig de la Bellacasa, Maria. 2015. "Making Time for Soil: Technoscientific Futurity and the Pace of Care." *Social Studies of Science* 45 (5): 691–716. https://doi.org/10 .1177/0306312715599851.

Puig de la Bellacasa, Maria. 2017. *Matters of Care: Speculative Ethics in More than Human Worlds*. Minneapolis: University of Minnesota Press.

Rigby, Kate, and Owain Jones. 2022. "Roadkill: Multispecies Mobility and Everyday Ecocide." In *Kin: Thinking with Deborah Bird Rose*, edited by Thom van Dooren and Matthew Chrulew, 112–134. Durham, NC: Duke University Press.

Riley, Crystal, Srikantan Jayasri, and Michael Gumert. 2015. "Results of a Nation-wide Census of the Long-Tailed Macaque (*Macaca Fascicularis*) Population of Singapore." *Raffles Bulletin of Zoology* 63 (October): 503–515.

Robbins, Paul. 2007. *Lawn People: How Grasses, Weeds, and Chemicals Make Us Who We Are*. Philadelphia: Temple University Press.

Robertson, Morgan M. 2004. "The Neoliberalization of Ecosystem Services: Wetland Mitigation Banking and Problems in Environmental Governance." *Geoforum* 35 (3): 361–373. https://doi.org/10.1016/j.geoforum.2003.06.002.

Robinson, Fiona. 2011. *The Ethics of Care: A Feminist Approach to Human Security*. Philadelphia: Temple University Press.

Rose, Deborah Bird. 2004. *Reports from a Wild Country: Ethics for Decolonisation*. Sydney: UNSW Press.

Rose, Deborah Bird. 2005. "The Rain Keeps Falling." *Cultural Studies Review* 11 (1). 122–127.

Rose, Deborah Bird. 2006. "What If the Angel of History Were a Dog?" *Cultural Studies Review* 12 (1): 67–78. https://doi.org/10.5130/csr.v12i1.3414.

Rose, Deborah Bird. 2013. "Slowly ~ Writing into the Anthropocene." *TEXT Special Issue* 17 (20): 1–14. https://doi.org/10.52086/001c.28826.

Rose, Deborah Bird, Thom Van Dooren, Matthew Chrulew, Stuart Cooke, Matthew Kearnes, and Emily O'Gorman. 2012. "Thinking through the Environment, Unsettling the Humanities." *Environmental Humanities* 1 (1): 1–5. https://doi.org/10.1215 /22011919-3609940.

Rose, Kira Alexandra. 2021. "Singapore's Liquid National Identity." *ISLE: Interdisciplinary Studies in Literature and Environment* 28 (3): 817–838. https://doi.org/10.1093 /isle/isaa094.

Rosenberg, Daniel K., Barry R. Noon, and E. Charles Meslow. 1997. "Biological Corridors: Form, Function, and Efficacy." *BioScience* 47 (10): 677–687. https://doi.org/10.2307/1313208.

Roy, Arundhati. 2020. "The Pandemic Is a Portal." *Financial Times*, April 4. https://www.ft.com/content/10d8f5e8-74eb-11ea-95fe-fcd274e920ca.

Sandilands, Catriona, Olga Cielemęcka, and Marianna Szczygielska. 2019. "Thinking the Feminist Vegetal Turn in the Shadow of Douglas-Firs: An Interview with Catriona Sandilands." *Catalyst: Feminism, Theory, Technoscience* 5 (2). https://doi.org/10.28968/cftt.v5i2.32863.

Schneider-Mayerson, Matthew. 2017. "Some Islands Will Rise: Singapore in the Anthropocene." *Resilience: A Journal of the Environmental Humanities* 4 (2–3): 166. https://doi.org/10.5250/resilience.4.2-3.0166.

Scott, James C. 1998. *Seeing Like a State: How Certain Schemes to Improve the Human Condition Have Failed*. New Haven, CT: Yale University Press.

Seah, May. 2016. "Singapore on the Map in Planet Earth II." *Today Online*. November 16. https://www.todayonline.com/entertainment/television/singapore-map-planet-earth-ii.

Senthilingam, Meera. 2015. "5 Ideas Every City Should Steal from Singapore." *CNN*, June 18.

Sha, John Chih Mun, Michael D. Gumert, Benjamin P. Y.-H. Lee, Agustin Fuentes, Subaraj Rajathurai, Sharon Chan, and Lisa Jones-Engel. 2009. "Status of the Long-Tailed Macaque Macaca Fascicularis in Singapore and Implications for Management." *Biodiversity and Conservation* 18 (11): 2909–2926. https://doi.org/10.1007/s10531-009-9616-4.

Sharifi, Ayyoob. 2016. "From Garden City to Eco-Urbanism: The Quest for Sustainable Neighborhood Development." *Sustainable Cities and Society* 20 (January): 1–16. https://doi.org/10.1016/j.scs.2015.09.002.

Shove, Elizabeth. 2018. "What Is Wrong with Energy Efficiency?" *Building Research & Information* 46 (7): 779–789. https://doi.org/10.1080/09613218.2017.1361746.

Siau, Ming En. 2017. "The Big Read: As Animal Encounters Hit the Headlines, a Divide Opens Up." *Today Online*, February 18.

Sim, Dewey. 2019. "Why Can't Southeast Asia Snuff Out Its Haze Problem for Good?" *South China Morning Post*, September 20. https://www.scmp.com/week-asia/health-environment/article/3029754/why-cant-southeast-asia-snuff-out-its-haze-problem.

Singapore Department of Statistics. 2019. "Population and Population Structure." Singapore Department of Statistics.

Singapore Government. 2016. "Questions on the Cross Island Line." Singapore Government.

Singapore Government. 2018. "Can the Prices in the 1962 Water Agreement Be Revised?" Singapore Government, July 9.

Singapore Government. 2021. *Singapore Green Plan 2030*. Singapore Government. https://www.greenplan.gov.sg/.

Singapore Press Holdings. 1963. "Prime Minister Lee Kuan Yew Planting a Sapling at Holland Circus during His Tour of Ulu Pandan Constituency." June 16. National Archives of Singapore.

Singapore Standards Council. 2019. *Singapore Standard: Specification for Organic Primary Produce. Enterprise Singapore*. Singapore: Enterprise Singapore.

Singh, Bryna. 2016. "Vertical Farms on the Rise in Land Scarce Singapore." *The Straits Times*, July 10.

Slater, Joanna. 2020. "In India, Life under Coronavirus Brings Blue Skies and Clean Air." *The Washington Post*, April 11.

Snell, Stuart. 2018. "World-First Major Desalination Field Study Finds Minimal Marine Impact." *UNSW Sydney Newsroom*, September 20.

Sofoulis, Zoë. 2005. "Big Water, Everyday Water: A Sociotechnical Perspective." *Continuum* 19 (4): 445–463. https://doi.org/10.1080/10304310500322685.

Soper, Kate. 1995. *What Is Nature?: Culture, Politics and the Non-Human*. Oxford and Cambridge, UK: Blackwell.

Spaargaren, Gert, and Arthur P. J. Mol. 1992. "Sociology, Environment, and Modernity: Ecological Modernization as a Theory of Social Change." *Society & Natural Resources* 5 (4): 323–344. https://doi.org/10.1080/08941929209380797.

Star, Susan Leigh. 1999. "The Ethnography of Infrastructure." *American Behavioral Scientist* 43 (3): 377–391.

Steele, Wendy, Ilan Wiesel, and Cecily Maller. 2019. "More-Than-Human Cities: Where the Wild Things Are." *Geoforum* 106 (November): 411–415. https://doi.org/10.1016/j.geoforum.2019.04.007.

Stengers, Isabelle. 2005. "The Cosmopolitical Proposal." Translated by Liz Carey-Libbrecht. In *Making Things Public: Atmospheres of Democracy*, edited by Bruno Latour and Peter Weibel, 994–1003. Cambridge, MA: MIT Press.

Stengers, Isabelle. 2010. *Cosmopolitics I*. Translated by Robert Bononno. Minneapolis: University of Minnesota Press.

Stengers, Isabelle. 2018. *A Manifesto for Slow Science*. Translated by Stephen Muecke. Cambridge, UK: Polity Press.

Stoetzer, Bettina. 2022. *Ruderal City: Ecologies of Migration, Race, and Urban Nature in Berlin*. Durham, NC: Duke University Press.

Strathern, Marilyn. 1992. *Reproducing the Future: Essays on Anthropology, Kinship, and the New Reproductive Technologies*. Manchester: Manchester University Press.

Subramanian, Samanth. 2017. "How Singapore Is Creating More Land for Itself." *The New York Times*, April 20. https://www.nytimes.com/2017/04/20/magazine/how-singapore-is-creating-more-land-for-itself.html.

Swyngedouw, Erik. 2007. "Water, Money and Power." *Socialist Register* 43: 195–212.

Swyngedouw, Erik. 2013. "Into the Sea: Desalination as Hydro-Social Fix in Spain." *Annals of the Association of American Geographers* 103 (2): 261–270. https://doi.org/10.1080/00045608.2013.754688.

Swyngedouw, Erik. 2015. *Liquid Power Contested Hydro-Modernities in Twentieth-Century Spain*. Cambridge, MA: MIT Press.

Swyngedouw, Erik, and Nikolas C. Heynen. 2003. "Urban Political Ecology, Justice and the Politics of Scale." *Antipode* 35 (5): 898–918.

Symons, Jonathan, and Rasmus Karlsson. 2018. "Ecomodernist Citizenship: Rethinking Political Obligations in a Climate-Changed World." *Citizenship Studies* 22 (7): 685–704.

Sze, Julie. 2015. *Fantasy Islands: Chinese Dreams and Ecological Fears in an Age of Climate Crisis*. Berkeley: University of California Press.

Szerszynski, Bronislaw. 2015. "Getting Hitched and Unhitched with the Ecomodernists." *Environmental Humanities* 7 (1): 239–244. https://doi.org/10.1215/22011919-3616443.

Szymanski, Erika Amethyst. 2018. "What Is the Terroir of Synthetic Yeast?" *Environmental Humanities* 10 (1): 40–62. https://doi.org/10.1215/22011919-4385462.

Tan, Adele. 2019. "Ask a Curator | Ng Teng Fong Roof Garden Commission: Charles Lim Yi Yong, SEA STATE 9: Proclamation Garden." *National Gallery Singapore*, August 16.

Tan, Adele. 2021. "Planting and Imaging Precarity in Charles Lim's SEASTATE 9: Proclamation Garden." *Antennae: The Journal of Nature in Visual Culture* 1 (54): 117–131.

Tan, Audrey. 2016. "Call for 'Zero Impact' for Cross Island MRT Line under MacRitchie Nature Reserve." *The Straits Times*, February 14. https://str.sg/UcsJ.

Tan, Audrey. 2017. "Will Desalination Affect Our Sea Water?" *The Straits Times*, July 8. https://str.sg/3QNU.

Tan, Audrey. 2019. "Singapore Land Went from Being Carbon Absorber to Emitter Due to Clearing of Forests: Report." *The Straits Times*, October 8. https://str.sg/J3wY.

Tan, Kenneth Paul. 2004. "Ethnic Representation on Singapore Film and Television." In *Beyond Rituals and Riots: Ethnic Pluralism and Social Cohesion in Singapore*, edited by An En Lai, 289–315. Singapore: Eastern Universities Press.

Tan, Pin Pin (Director). 2001. *Moving House*. Video File.

Tan, Pin Pin. 2012. "Interview: Tan Pin Pin." *Cinémathèque Quarterly*.

Tan, Puay Yok, and Abdul Rahim bin Abdul Hamid. 2014. "Urban Ecological Research in Singapore and Its Relevance to the Advancement of Urban Ecology and Sustainability." *Landscape and Urban Planning* 125 (May): 271–289. https://doi.org /10.1016/j.landurbplan.2014.01.019.

Tan, Puay Yok, James Wang, and Angelia Sia. 2013. "Perspectives on Five Decades of the Urban Greening of Singapore." *Cities* 32 (June): 24–32. https://doi.org/10.1016/j .cities.2013.02.001.

Tan, Yong Soon, Tung Jean Lee, and Karen Tan. 2016. "Cleaning the Land and Rivers." In *50 Years of Environment Singapore's Journey towards Environmental Sustainability*, edited by Yong Soon Tan, 15–44. Singapore: World Scientific.

Tanpinar, Ahmet Hamdi. 2014. *The Time Regulation Institute*. (Original work published as Saatleri Ayarlama Enstitiüsü in 1962). Translated by Maureen Freely and Alexander Dawe. London: Penguin.

Tay, Lai Hock. 2012. Interview with Tay Lai Hock. Profiled by Lim Zi Song. *We Are Singapore*.

Tay, Lai Hock. 2014. Interview with Tay Lai Hock. "Balik Kampong" Interview by F. Toh. *Archifest*.

Tay, Lai Hock. 2015. Interview with Bridgette See. "Can We Nurture Kampung Spirit in Singapore?" *Challenge*, June 30.

Tay, Ywee Chieh, Daniel Jia Jun Ng, Jun Bin Loo, Danwei Huang, Yixiong Cai, Darren Chong Jinn Yeo, and Rudolf Meier. 2018. "Roads to Isolation: Similar Genomic History Patterns in Two Species of Freshwater Crabs with Contrasting Environmental Tolerances and Range Sizes." *Ecology and Evolution* 8 (9): 4657–4668. https://doi.org /10.1002/ece3.4017.

Tepper, Laura. 2011. "Road Ecology: Wildlife Habitat and Highway Design." *Places Journal*. September. https://doi.org/10.22269/110922.

Thiagarajah, Jharyathri, Shermaine K. M. Wong, Daniel R. Richards, and Daniel A. Friess. 2015. "Historical and Contemporary Cultural Ecosystem Service Values in the Rapidly Urbanizing City State of Singapore." *Ambio* 44 (7): 666–677. https://doi.org /10.1007/s13280-015-0647-7.

Thulaja, Naidu Ratnala. n.d. "Gambier." *Singapore Infopedia, National Library Board*.

Timm, Stephanie N., and Brian M. Deal. 2018. "Understanding the Behavioral Influences behind Singapore's Water Management Strategies." *Journal of Environmental Planning and Management* 61 (10): 1654–1673. https://doi.org/10.1080/09640568 .2017.1369941.

Toh, Wen Li. 2017. "Replacing Lost Trees." *The Straits Times*, March 19. https://str.sg /3bhj.

Tortajada, Cecilia, Yugal Kishore Joshi, and Asit K. Biswas. 2013. *The Singapore Water Story: Sustainable Development in an Urban City-State*. London: Routledge.

Transsolar. 2013. Gardens by the Bay Conservatories, Singapore. Transsolar Klima Engineering.

Tronto, Joan. 1993. *Moral Boundaries: A Political Argument for an Ethics of Care*. New York: Routledge.

Tronto, Joan. 2005. "Care as the Work of: A Modest Proposal." In *Women and Citizenship*, edited by Marilyn Friedman, 130–145. New York: Oxford University Press.

Tsing, Anna Lowenhaupt. 2015. *The Mushroom at the End of the World: On the Possibility of Life in Capitalist Ruins*. Princeton, NJ: Princeton University Press.

Tuan, Yi-Fu. 1977. *Space and Place: The Perspective of Experience*. London: Edward Arnold.

Turnbull, C. M. 2009. *A History of Modern Singapore 1819–2005*. Singapore: NUS Press.

UN-Habitat. 2012. "Urban Patterns for a Green Economy: Working with Nature." *UN-Habitat*.

United Nations. 2011. *The Global Social Crisis: Report on the World Social Situation 2011*. New York: United Nations Publications.

Urban Redevelopment Authority. 2014. "Introduction: About Master Plan 2014." *Urban Redevelopment Authority*. Singapore.

Urban Redevelopment Authority. 2017. "Updates to the Landscaping for Urban Spaces and High-Rises (LUSH) Programme: LUSH 3.0." *Urban Redevelopment Authority*. Singapore.

Urban Redevelopment Authority. 2019. "Draft Master Plan 2019—Proposals for an Inclusive, Sustainable and Resilient City." *Urban Redevelopment Authority*. 27 March. Singapore.

Usher, Mark. 2018. "Conduct of Conduits: Engineering, Desire and Government through the Enclosure and Exposure of Urban Water." *International Journal of Urban and Regional Research* 42 (2): 315–333. https://doi.org/10.1111/1468-2427.12524.

Van Der Windt, Henny J., and J. A. A. Swart. 2008. "Ecological Corridors, Connecting Science and Politics: The Case of the Green River in the Netherlands." *Journal of Applied Ecology* 45 (1): 124–132. https://doi.org/10.1111/j.1365-2664.2007.01404.x.

van Dooren, Thom. 2014. *Flight Ways: Life and Loss at the Edge of Extinction*. New York: Columbia University Press.

van Dooren, Thom. 2015. "A Day with Crows—Rarity, Nativity and the Violent-Care of Conservation." *Animal Studies Journal* 4 (2): 1–28.

van Dooren, Thom. 2017. "Spectral Crows in Hawai'i: Conservation and the Work of Inheritance." In *Extinction Studies: Stories of Time, Death, and Generations*, edited by Deborah Bird Rose, Thom van Dooren, and Matthew Chrulew, 187–216. New York: Columbia University Press.

van Dooren, Thom. 2019. *The Wake of Crows: Living and Dying in Shared Worlds*. New York: Columbia University Press.

van Dooren, Thom. 2020a. "Story (Telling)." *Swamphen: A Journal of Cultural Ecology* 7. https://doi.org/10.60162/swamphen.7.14359.

van Dooren, Thom. 2020b. "Pangolins And Pandemics: The Real Source Of This Crisis Is Human, Not Animal." *New Matilda*. March 22.

van Dooren, Thom, Eben Kirksey, and Ursula Münster. 2016. "Multispecies Studies: Cultivating Arts of Attentiveness." *Environmental Humanities* 8 (1): 1–23. https://doi.org/10.1215/22011919-3527695.

van Dooren, Thom, and Deborah Bird Rose. 2016 "Lively Ethography: Storying Animist Worlds." *Environmental Humanities* 8 (1): 77–94. https://doi.org/10.1215/22011919-3527731.

Vanham, Davy. 2011. "How Much Water Do We Really Use? A Case Study of the City State of Singapore." *Water Science and Technology: Water Supply* 11 (2): 219–228. https://doi.org/10.2166/ws.2011.043.

Vartan, Starre. 2019. "How Wildlife Bridges over Highways Make Animals—and People—Safer." *National Geographic*, April 16.

Vincent, Lenouvel, Lafforgue Michel, Chevauché Catherine, and Rhétoré Pauline. 2014. "The Energy Cost of Water Independence: The Case of Singapore." *Water Science and Technology: A Journal of the International Association on Water Pollution Research* 70 (5): 787–794. https://doi.org/10.2166/wst.2014.290.

Virilio, Paul. 2001. *Virilio Live: Selected Interviews*, edited by John Armitage. London: SAGE Publications. https://doi.org/10.4135/9781446220306.

Virilio, Paul. 2006. *Speed and Politics*. Translated by Mark Polizzotti. Los Angeles: Semiotext(e).

Wakefield, S. 2020. "Making Nature into Infrastructure: The Construction of Oysters as a Risk Management Solution in New York City." *Environment and Planning E: Nature and Space* 3 (3): 761–785. https://doi.org/10.1177/2514848619887461.

Wallace, Scott. 2020. "Disaster Looms for Indigenous Amazon Tribes as COVID-19 Cases Multiply." *National Geographic*.

Wang, James Wei, Choon Hock Poh, Chloe Yi Ting Tan, Vivien Naomi Lee, Anuj Jain, and Edward L. Webb. 2017. "Building Biodiversity: Drivers of Bird and Butterfly Diversity on Tropical Urban Roof Gardens." *Ecosphere* 8 (9): e01905. https://doi.org/10.1002/ecs2.1905.

Wang, Jamie. 2021. "Urban." *A-Z of Shadow Places Concepts*. https://www.shadowplaces.net/concepts.

Weale, Albert. 1992. *The New Politics of Pollution*. Manchester: Manchester University Press.

Wee, Kellynn, Theodora Lam, and Brenda S. A. Yeoh. 2022. "Migrant Construction Workers in Singapore: An Introduction." In *Migrant Workers in Singapore: Lives and Labour in A Transient Migration Regime*, edited by Brenda S. A. Yeoh, Kellynn Wee, and Theodora Lam, xiii–xlix. Singapore: World Scientific Publishing Company.

West, Harry G., and Nuno Domingos. 2012. "Gourmandizing Poverty Food: The Serpa Cheese Slow Food Presidium." *Journal of Agrarian Change* 12 (1): 120–143. https://doi.org/10.1111/j.1471-0366.2011.00335.x.

Weston, Kath. 2017. *Animate Planet: Making Visceral Sense of Living in a High-Tech Ecologically Damaged World*. Durham, NC: Duke University Press.

Whatmore, Sarah. 2006. "Materialist Returns: Practising Cultural Geography in and for a More-Than-Human World." *Cultural Geographies* 13 (4): 600–609. https://doi .org/10.1191/1474474006cgj377oa.

Whitehead, Richard. 2019. "Singapore Farmers Hope Government Gives Clarity on the Future of Their Lands." *Dairy Reporter*, March 27.

Whitington, Jerome. 2016. "Modernist Infrastructure and the Vital Systems Security of Water: Singapore's Pluripotent Climate Futures." *Public Culture* 28 (2 (79)): 415–441. https://doi.org/10.1215/08992363-3427511.

Williams, Joe, and Erik Swyngedouw, eds. 2018. *Tapping the Oceans: Seawater Desalination and the Political Ecology of Water*. Cheltenham and Northampton: Edward Elgar Publishing.

Williams, Miriam J. 2020. "The Possibility of Care-Full Cities." *Cities* 98:1–7. https:// doi.org/10.1016/j.cities.2019.102591.

Williams, Nicholas S. G., Jeremy Lundholm, and J. Scott MacIvor. 2014. "Do Green Roofs Help Urban Biodiversity Conservation?" *Journal of Applied Ecology* 51 (6): 1643–1649. https://doi.org/10.1111/1365-2664.12333.

Witt, Emily. 2022. "An Urban Wildlife Bridge Is Coming to California." *The New Yorker*, May 17.

Wolch, Jennifer. 2002. "Anima Urbis." *Progress in Human Geography* 26 (6): 721–742. https://doi.org/10.1191/0309132502ph400oa.

Wolch, Jennifer. 2007. "Green Urban Worlds." *Annals of the Association of American Geographers* 97 (2): 373–384. https://doi.org/10.1111/j.1467-8306.2007.00543.x.

Wolch, Jennifer R., Jason Byrne, and Joshua P. Newell. 2014. "Urban Green Space, Public Health, and Environmental Justice: The Challenge of Making Cities 'Just Green Enough.'" *Landscape and Urban Planning* 125 (May): 234–244. https://doi.org /10.1016/j.landurbplan.2014.01.017.

Wong, Catherine Mei Ling. 2012. "The Developmental State in Ecological Modernization and the Politics of Environmental Framings: The Case of Singapore and Implications for East Asia." *Nature and Culture* 7 (1): 95–119. https://doi.org/10.3167 /nc.2012.070106.

Wong, May Ee. 2022. "Visualising 'Asia as Future' through Speculative Southeast Asian Aesthetic Urban Futures." *ARIscope—National University of Singapore*, August 25.

Wong, Pei Ting. 2019. "Shocking that Singaporeans Ask for Wildlife to Be 'Returned' to the Zoo, Says Jane Goodall." *Today*, November 27.

Wong, Poh Poh. 2018. "Coastal Protection Measures—Case of Small Island Developing States to Address Sea-Level Rise." *Asian Journal of Environment & Ecology* 6 (3): 1–14. https://doi.org/10.9734/ajee/2018/41019.

The World Bank. 2020. "Global Economic Prospects, June 2020." *The World Bank*, June.

The World Bank. 2021. "Population Density People per Sq. Km of Land Area." *The World Bank*.

World Commission on Environment and Development (WCED). 1987. *Report of the World Commission on Environment and Development: Our Common Future*. Oxford: Oxford University Press.

World Economic Forum. 2015. *Global Risks 2015,* 10th ed. Geneva: World Economic Forum. https://www3.weforum.org/docs/WEF_Global_Risks_2015_Report15.pdf.

World Health Organization. 2011. *Safe Drinking-Water from Desalination*. Geneva: WHO Press.

World Health Organization @WHO. 2020. "#Singapore Is Leaving No Stone Unturned, Testing Every Case of Influenza-like Illness and Pneumonia. So Far They Have Not Found Evidence of #COVID19 Community Transmission." *Twitter*, February 19. https://twitter.com/WHO/status/1229797987888615425.

Worster, Donald. 1992. *Rivers of Empire: Water, Aridity and the Growth of the American West*. Oxford: Oxford University Press.

Wright, Katherine. 2012. "Pining for the Present: Ecological Remembrance and Healing in the Armidale State Forest." *Environmental Philosophy* 9 (1): 109–126. https://doi.org/10.5840/envirophil2012917.

Yahoo. 2013. "4,000 Turn up at Speakers' Corner for Population White Paper Protest." *Yahoo!*, February 16.

Yang, Hui, Peidong Yang, and Shaohua Zhan. 2017. "Immigration, Population, and Foreign Workforce in Singapore: An Overview of Trends, Policies, and Issues." *HSSE* 6 (1): 10–25.

Yang, Yang, Peitong Zhang, and Shaoquan Ni. 2014. "Assessment of the Impacts of Urban Rail Transit on Metropolitan Regions Using System Dynamics Model." *Transportation Research Procedia* 4:521–534. https://doi.org/10.1016/j.trpro.2014.11.040.

Yap, Jo Leen, Nik Fadzly Nik Rosely, Mazrul Mahadzir, Mark Louis Benedict, Vikneswaran Muniandy, and Nadine Ruppert. 2022. "'Ah Lai's Crossing'—Malaysia's First Artificial Road Canopy Bridge to Facilitate Safer Arboreal Wildlife Crossings." *Folia Primatologica* 93 (3–6): 255–269. https://doi.org/10.1163/14219980-20211105.

Yea, Sallie. 2020. "This Is Why Singapore's Coronavirus Cases Are Growing: A Look inside the Dismal Living Conditions of Migrant Workers." *The Conversation*, April 29. https://theconversation.com/this-is-why-singapores-coronavirus-cases-are-growing-a-look-inside-the-dismal-living-conditions-of-migrant-workers-136959.

Yee, Alex Thiam Koon, Kwek Yan Chong, Louise Neo, and Hugh T. W. Tan. 2016. "Updating the Classification System for the Secondary Forests of Singapore." *Raffles Bulletin of Zoology* 32 (May): 11–21.

Yeo, Jun-Han, and Harvey Neo. 2010. "Monkey Business: Human–Animal Conflicts in Urban Singapore." *Social & Cultural Geography* 11 (7): 681–699. https://doi.org/10.1080/14649365.2010.508565.

Yeoh, Brenda S. A. 2006. "Bifurcated Labour: The Unequal Incorporation of Transmigrants in Singapore." *Tijdschrift voor Economische en Sociale Geografie* 97 (1): 26–37.

Yeoh, Brenda S. A., and Lily Kong. 2012. "Singapore's Chinatown: Nation Building and Heritage Tourism in a Multiracial City." *Localities* 2:117–159.

Yeoh, Brenda S. A., and Maria Andrea Soco. 2014. "The Cosmopolis and the Migrant Domestic Worker." *Cultural Geographies* 21 (2): 171–187. https://doi.org/10.1177/1474474014520899.

York, Richard, and Julius Alexander McGee. 2016. "Understanding the Jevons Paradox." *Environmental Sociology* 2 (1): 77–87. https://doi.org/10.1080/23251042.2015.1106060.

York, Richard, and Eugene A. Rosa. 2003. "Key Challenges to Ecological Modernization Theory: Institutional Efficacy, Case Study Evidence, Units of Analysis, and the Pace of Eco-Efficiency." *Organization & Environment* 16 (3): 273–288. https://doi.org/10.1177/1086026603256299.

Yuen, Belinda. 1996. "Creating the Garden City: The Singapore Experience." *Urban Studies* 33 (6): 955–970. https://doi.org/10.1080/00420989650011681.

Yuen, Belinda. 2005. "Searching for Place Identity in Singapore." *Habitat International* 29 (2): 197–214. https://doi.org/10.1016/j.habitatint.2003.07.002.

Yuen, Belinda. 2011. "Urban Planning in Southeast Asia: Perspective from Singapore." *Town Planning Review* 82 (2): 145–168. https://doi.org/10.3828/tpr.2011.12.

Zulkifli, Masagos. 2019a. Committee of Supply Debate 2019-Mr Masagos Zulkifli, Minister for the Environment and Water Resources. *Ministry of Sustainability and the Environment*. March 7.

Zulkifli, Masagos. 2019b. "Speech by Mr. Masagos Zulkifli, Minister for the Environment and Water Resources, at the Sky Greens SS 632 Certification Ceremony, on 11 June 2019." National Archives of Singapore.

INDEX

Photos and illustrations are indicated by italicized page numbers.

Urban and Industrial Environments

Series editors: Robert Gottlieb, Henry R. Luce Professor of Urban and Environmental Policy, Occidental College

Bhavna Shamasunder, Associate Professor of Urban and Environmental Policy, Occidental College

Maureen Smith, *The U.S. Paper Industry and Sustainable Production: An Argument for Restructuring*

Keith Pezzoli, *Human Settlements and Planning for Ecological Sustainability: The Case of Mexico City*

Sarah Hammond Creighton, *Greening the Ivory Tower: Improving the Environmental Track Record of Universities, Colleges, and Other Institutions*

Jan Mazurek, *Making Microchips: Policy, Globalization, and Economic Restructuring in the Semiconductor Industry*

William A. Shutkin, *The Land That Could Be: Environmentalism and Democracy in the Twenty-First Century*

Richard Hofrichter, ed., *Reclaiming the Environmental Debate: The Politics of Health in a Toxic Culture*

Robert Gottlieb, *Environmentalism Unbound: Exploring New Pathways for Change*

Kenneth Geiser, *Materials Matter: Toward a Sustainable Materials Policy*

Thomas D. Beamish, *Silent Spill: The Organization of an Industrial Crisis*

Matthew Gandy, *Concrete and Clay: Reworking Nature in New York City*

David Naguib Pellow, *Garbage Wars: The Struggle for Environmental Justice in Chicago*

Julian Agyeman, Robert D. Bullard, and Bob Evans, eds., *Just Sustainabilities: Development in an Unequal World*

Barbara L. Allen, *Uneasy Alchemy: Citizens and Experts in Louisiana's Chemical Corridor Disputes*

Dara O'Rourke, *Community-Driven Regulation: Balancing Development and the Environment in Vietnam*

Brian K. Obach, *Labor and the Environmental Movement: The Quest for Common Ground*

Peggy F. Barlett and Geoffrey W. Chase, eds., *Sustainability on Campus: Stories and Strategies for Change*

Steve Lerner, *Diamond: A Struggle for Environmental Justice in Louisiana's Chemical Corridor*

Jason Corburn, *Street Science: Community Knowledge and Environmental Health Justice*

Peggy F. Barlett, ed., *Urban Place: Reconnecting with the Natural World*

David Naguib Pellow and Robert J. Brulle, eds., *Power, Justice, and the Environment: A Critical Appraisal of the Environmental Justice Movement*

Eran Ben-Joseph, *The Code of the City: Standards and the Hidden Language of Place Making*

Nancy J. Myers and Carolyn Raffensperger, eds., *Precautionary Tools for Reshaping Environmental Policy*

Kelly Sims Gallagher, *China Shifts Gears: Automakers, Oil, Pollution, and Development*

Kerry H. Whiteside, *Precautionary Politics: Principle and Practice in Confronting Environmental Risk*

Benjamin Pauli, *Flint Fights Back: Environmental Justice and Democracy in the Flint Water Crisis*

Karen Chapple and Anastasia Loukaitou-Sideris, *Transit-Oriented Displacement or Community Dividends? Understanding the Effects of Smarter Growth on Communities*

Henrik Ernstson and Sverker Sörlin, eds., *Grounding Urban Natures: Histories and Futures of Urban Ecologies*

Katrina Smith Korfmacher, *Bridging the Silos: Collaborating for Environment, Health, and Justice in Urban Communities*

Jill Lindsey Harrison, *From the Inside Out: The Fight for Environmental Justice within Government Agencies*

Anastasia Loukaitou-Sideris, Dana Cuff, Todd Presner, Maite Zubiaurre, and Jonathan Jae-an Crisman, *Urban Humanities: New Practices for Reimagining the City*

Govind Gopakumar, *Installing Automobility: Emerging Politics of Mobility and Streets in Indian Cities*

Amelia Thorpe, *Everyday Ownership: PARK(ing) Day and the Practice of Property*

Tridib Banerjee, *In the Images of Development: City Design in the Global South*

Ralph Buehler and John Pucher, eds., *Cycling for Sustainable Cities*

Casey J. Dawkins, *Just Housing: The Moral Foundations of American Housing Policy*

Kian Goh, *Form and Flow: The Spatial Politics of Urban Resilience and Climate Justice*

Kian Goh, Anastasia Loukaitou-Sideris, and Vinit Mukhija, eds., *Just Urban Design: The Struggle for a Public City*

Sheila R. Foster and Christian Iaione, *Co-Cities: Innovative Transitions toward Just and Self-Sustaining Communities*

Vinit Mukhija, *Remaking the American Dream: The Informal and Formal Transformation of Single-Family Housing Cities*

Cindy McCulligh, *Sewer of Progress: Corporations, Institutionalized Corruption, and the Struggle for the Santiago River*

Susan Handy, *Shifting Gears: The History and Future of American Transportation*

Manisha Anantharaman, *Communal Sustainability: Gleaning Justice in Bengaluru's Discards*

Zachary B. Lamb and Lawrence J. Vale, *The Equitably Resilient City: Struggles and Solidarities in the Face of Climate Crisis*

J. Mijin Cha, *A Just Transition for All: Workers, Communities, and a Carbon-Free Future*

Jamie Wang, *Reimagining the More-Than-Human City: Stories from Singapore*